The MIS and LAN Managers Guide to Advanced Telecommunications

Leo A. Wrobel

IEEE
COMPUTER
SOCIETY

Los Alamitos, California

Washington • Brussels • Tokyo

Library of Congress Cataloging-in-Publication Data

Wrobel, Leo A. (Leo Anthony)
  The MIS and LAN managers guide to advanced telecommunications /
Leo A. Wrobel.
    p.   cm.
  ISBN 0-7695-0099-4
  1. Local area networks (Computer networks) 2. Wide area networks
(Computer networks)   I. Title.

  TK5105.7 .W76 1999
  004.6'8 — dc21

                                                    99-046044
                                                        CIP

IEEE Computer Society Press Order Number BP00099
Library of Congress Number 99-046044
ISBN 0-7695-0099-4

Additional copies may be ordered from:

IEEE Computer Society Press	IEEE Service Center	IEEE Computer Society
Customer Service Center	445 Hoes Lane	Watanabe Building
10662 Los Vaqueros Circle	P.O. Box 1331	1-4-2 Minami-Aoyama
P.O. Box 3014	Piscataway, NJ 08855-1331	Minato-ku, Tokyo 107-0062
Los Alamitos, CA 90720-1314	Tel: +1-732-981-0060	JAPAN
Tel: +1-714-821-8380	Fax: +1-732-981-9667	Tel: +81-3-3408-3118
Fax: +1-714-821-4641	mis.custserv@computer.org	Fax: +81-3-3408-3553
cs.books@computer.org		tokyo.ofc@computer.org

Executive Director and Chief Executive Officer: T. Michael Elliott
Manager of Production, CS Press: Deborah Plummer
Advertising/Promotions: Tom Fink
Production Editor: Denise Hurst
Printed in the United States of America

Acknowledgments

- Our thanks to Premiere Network Services Inc. for their extensive contributions and examples. Premiere is a **Designer Carrier**^{TM} catering to the needs of "Fortune" users. More information about Premiere is available at www.rewireit.com or 1-888-REWIREIT. Special thanks to the following staff members in alphabetical order:

  Archie M. Croom, VP Engineering
  Mary Beth Granzow, Administrative Assistant
  Sharon Ford, Assistant to the President
  Jacquetta L. Peace, Operations Manager
  Chavvon Smith, Sr. Financial Analyst
  Sarah Wrobel, Graphics Coordinator

- Portions of this book were taken from articles written for *Technical Support* magazine, published by Technical Enterprises, Inc. for NaSPA, Inc. (the Association for Corporate Computing Technical Professionals) as well as from the Artech House publications listed in the Biography at the end of this book.

- The Author also wishes to thank Eddie M. Pope, Esq. of the Austin-based law firm of Clark, Thomas and Winters for his contributions. Eddie can be reached at 1-512-495-8896.

- Our thanks to Bud Bates, and TCI International Consulting for his contribution. Information about TCI International Consulting is available at www.tcic.com.

Contents

Preface   **xix**

1   First, the Basics   **1**

1.1   LANs, WANs, What Is the Difference? 1

1.2   So This Is Basic? 1

     1.2.1   What Do Mainframes Do? 2

     1.2.2   What Is a Front-End Processor? 2

     1.2.3   What Are LANS? 4

     1.2.4   How Do We Connect LANS? 5

     1.2.5   What Is Frame Relay? 5

     1.2.6   What Do LAN Switches Do? 6

     1.2.7   What Is ATM? 7

     1.2.8   What Is SONET? 7

     1.2.9   What Is Fiber? 8

     1.2.10   What Is the Internet? 8

     1.2.11   What Is a PBX? 9

     1.2.12   What Is a Channel Bank? 9

     1.2.13   What Is a T1? 9

     1.2.14   What Are Private Lines? 10

     1.2.15   What Is ISDN? 10

1.3   How Does Data Flow Through This Network? 11

1.4   PBX to PBX 15

1.5   Token Ring User to Local Ethernet User 15

1.6   Local Ethernet User to Remote Terminal User via Private Line 20

1.7   Local Ethernet User to Remote Ethernet User via Frame Relay 20

1.8   Voice Over Frame Relay 22

1.9   SONET and ATM 22

2   LAN Basics   **25**

2.1   What Is a LAN? 25

2.2   LAN Operations "Refresher" 25

     2.2.1   Limited Geographic Coverage 27

          2.2.2   LANs Are Customer Owned   29
          2.2.3   LANs Are Designed for Data   29
    2.3   Fast LANs   31
    2.4   How Fast Can You Run on a Wire?   32
    2.5   Are LANs Isochronous?   32
    2.6   Fast Ethernet   34
    2.7   100 VG/AnyLAN   36
    2.8   ISO-Ethernet (IEEE 802.9a)   37
    2.9   FDDI (Fiber Distributed Data Interface)   39
    2.10 Gigabit Ethernet   39
    2.11 ATM   41
    2.12 LAN Topologies   44
    2.13 Multimedia Applications   44
    2.14 Engineering "Killer Apps" Through Seamless Solutions   47
    2.15 Summary   51

3   **WAN Services Overview**   **53**
    3.1   Tying It Together   53
    3.2   The WAN Environment   53
    3.3   What Are Modems?   55
    3.4   Modem Performance Impairments   58
    3.5   56 Kbps Modem—Poor Man's ISDN   59

4   **LAN/WAN Interconnection Basics**   **63**
    4.1   Growing Killer Apps   63
    4.2   Developing a Strategic Plan   64
    4.3   What Is the Mission?   64
    4.4   Examples of Killer Apps   65
    4.5   The Key to Success—The Network   67
    4.6   Vendor-Managed Relationship   69
    4.7   Why Is Network Management Important?   73
    4.8   Choosing the Right Network Management System   73
    4.9   Selecting the Right System   74
    4.10 Packaging and Selling Network Management   74
    4.11 Defining the Components of a Network Management System   75
    4.12 The Purpose of a Network Management System   76
    4.13 Managing Network Diversity and Change   76
    4.14 Desirable Components of a Network Management System   77
          4.14.1 Fault Management   77
          4.14.2 Operations Management   77

        4.14.3 Performance Management 77
        4.14.4 Configuration Management 78
        4.14.5 Change Management 78
        4.14.6 Security Management 78
        4.14.7 Accounting Management 78
        4.14.8 Availability of a "Shrink-Wrapped" Solution 78
    4.15 Summary 79

5 High Speed WAN Connectivity (T1, T3, SONET)   81
    5.1   Buying the "Big Stuff" 81
    5.2   The "Old Stuff"—T1 82
    5.3   How We Have Done Business 82
    5.4   Satisfying Today's Users with Tomorrow's Technology 85
    5.5   What Do Large Users Want? 85
    5.6   What Is Great About SONET? 85
    5.7   "The Old Days" 86
    5.8   Today's Technology 86
    5.9   What Is State of the Art? 86
    5.10 What Is Out There Now? 86
    5.11 How Do We Position? What Is the Migration Path? 87
    5.12 What Is a T1? 87
    5.13 T1 Protocols 89
    5.14 B8ZS 89
    5.15 ESF 90
    5.16 Fault Conditions 94
    5.17 T3 94
    5.18 M13 Headaches and Shortcomings 94
    5.19 Ski Lift Analogy 95
    5.20 The DS3 Signal 97
    5.21 C-Bit Parity 97
    5.22 SYNTRAN 98
    5.23 SONET!—A "Smart" Network 99
    5.24 SONET Framed Signal (STS-1) 99
    5.25 SONET Link Architecture 100
    5.26 Other SONET Points 100
        5.26.1 SONET Is Inefficient—So What? 100
        5.26.2 Plenty of Room for a Brain 101
        5.26.3 Smarts in all the Boxes 101
        5.26.4 No Hassles with CRC Checking 101
        5.26.5 Convenient Connection to SDH 101

5.27 What About the Fiber Itself? 102
    5.27.1 Multimode 102
    5.27.2 Single Mode 102
5.28 Summary 102

6 Advanced Techniques for High Speed WAN Connectivity   105
6.1 "Buck Rogers" Telecom Applications 105
6.2 "Fortune" Company Checklist for Success 109
6.3 SONET Technology 113
    6.3.1 Fault Tolerant Ring Topology 113
    6.3.2 Allows 10, 16, and 100 Mbps Circuit Sizes 113
    6.3.3 Drop and Insert Capability 113
    6.3.4 Advanced Diagnostics 113
    6.3.5 Ability to Burst 114
6.4 How Do We Begin? 114
6.5 Using Advanced Networks to Manage Client-Server Computing 114
6.6 New Operating Modes Using SONET 116
6.7 Superior Network Management, Security, and Control 116
6.8 Greater Reliability and Network Management 118
6.9 How Does One Get SONET? 118
6.10 Bring Your Wallet 118
6.11 The Grand Strategy 119
6.12 Other Important Points 119
6.13 The Promise of ATM 121
6.14 "Rewiring" Frame Relay—One Promise of ATM 121

7 Supporting Small Offices, Nomadic Users, and Telecommuters   131
7.1 Introduction 132
7.2 Cost Considerations in Virtual Offices 132
7.3 Getting Employees the Tools They Need at Home 135
7.4 What About Communications? 137
7.5 Which Jobs are Best Suited for Nomadic Operations or Home Officing? . . . 137
7.6 Enabling Technologies for Telecommuting and Remote Access 140
    7.6.1 Dial-Up Data 140
    7.6.2 Modem Technology 140
    7.6.3 Modem Manufacturers 141
    7.6.4 Modem Standards and Speed 141
    7.6.5 Modem Configurations and Related Settings 142
    7.6.6 Other Issues Affecting Modem Communications . . 143
    7.6.7 What About the Phone Line? 143

	7.6.8	Consider an Internal Modem	144
	7.6.9	Don't Overlook the Obvious!	144
	7.6.10	Providing Internet Access	144
7.7	The Future is Digital		145
	7.7.1	T1 Services	146
7.8	Users Will Drive the Future		146
7.9	ISDN: Innovations Subscribers Don't *Do* Need!		147
	7.9.1	Check Availability in Your Area	148
	7.9.2	Select a Service Plan	148
	7.9.3	Select Required Features	149
	7.9.4	Select Your ISDN Equipment	149
	7.9.5	Plan for Power Backup!	150
	7.9.6	Place Your Order/Monitor Progress	151
	7.9.7	Select an ISDN Capable ISP (Internet Service Provider)	151
7.10	ISDN and Remote Access Equipment Examples		151
7.11	ISDN to the Internet		152
7.12	ISDN! How to Do It		153
7.13	Summary		159

8 Emerging Competitive Alternatives — **161**

8.1	The Telecommunications Reform Act of 1996 and its Implications on the SOHO Market	161
8.2	First of the "Emerging Technologies"—ADSL	162
8.3	DSL Equipment	163
8.4	DSL Speeds and Distances	163
8.5	VDSL	165
8.6	Cable in the SOHO Market	165
8.7	ADSL as a Competitive Threat to Cable	165
8.8	Cable-Based High Speed Telecom	166
8.9	Equipment Your Users Will Need	166
8.10	Another Cable Threat—Direct Satellite Service	167
8.11	Cable Engineering Pitfalls	167
8.12	Standards Concerns	168
8.13	Crafting a Strategy—How Cable Should Respond	168
8.14	Cable Market Opportunities	169
8.15	Traditional Telephone Service Resale (Two Dollar Window Services)	169
8.16	Enhanced SOHO Services (Five Dollar Window Services)	171
8.17	Competing on a Budget	172
8.18	The "Fortune" Level Customer ($100 Window)	173
	8.18.1 Fortune Users Have Insatiable Demands for Capacity	173

      8.18.2 Fortune Users Need Help Desk Support 173
      8.18.3 Fortune Users Need Data System Backups and Archival 174
      8.18.4 Fortune Users Need Network Management 174
  8.19 Selling Network Management to Fortune Users 174
  8.20 "Hybrid" Solutions 175
  8.21 Summary 175

9 Exploiting Telecom Reform in Network Procurements   179
  9.1   New Technologies = New Advantages 179
  9.2   Comparing Physical and Electronic Distribution Mediums   180
  9.3   Communicating the Vision 181
  9.4   Selling the Concept Internally 183
  9.5   Kicking Things Off 183
  9.6   Doing a Custom Tariff—Learning the Rules 184
      9.6.1   Learn the Language 184
      9.6.2   Converting "Feature Packages" into a Custom Tariff 186
  9.7   Influencing Telecommunications Policy Makers 186
  9.8   First, Talk to Each Other! 188
  9.9   A Little Bit of History 188
  9.10 Federal Law and the FCC 190
  9.11 State Law and the PUCs 191
  9.12 Local Law and the Cities 191
  9.13 Regulation Varies by Technology 192
  9.14 The Federal Telecommunications Act of 1996—A Lawyer's Perspective . . 192
  9.15 There Was *Supposed* to Be a War! 193
  9.16 The Act Is a Great Success 193
  9.17 Outline of the Basic Provisions 195
  9.18 The One Thing you Can Predict—Unpredictibility 195
  9.19 Watch the Sidelines—New Players May Be Coming! 195

10 Summary and Conclusions   197
  10.1 Becoming an "Activist" Telecommunications User 197
  10.2 Exploiting Telecom Reform—Defining the Project 198
      10.2.1 The Incumbent Local Exchange Carriers (ILECs) 198
      10.2.2 The Competitive Local Exchange Carriers (CLECs) 199
      10.2.3 The InterExchange Carriers (IXCs) 199
      10.2.4 The Policymakers 199
      10.2.5 The Internet Service Providers (ISPs) 201
      10.2.6 The "Activist" Fortune User 201
      10.2.7 So, What Does This Mean? 202

10.3 Who Decides What You Pay? 202

      10.3.1 Who Participates in Rate Cases? 205

      10.3.2 What Can the Fortune User Do to Influence Rates? 206

10.4 What Is Unbundling? 206

10.5 Exploiting Today, Preparing for Tomorrow 210

      10.5.1 Phase I—Concept Validation 210

      10.5.2 Phase II—Specifications, Configurations, and Final Plans 210

      10.5.3 Phase III—Regulatory Support 210

      10.5.4 Phase IV—Exploiting Telecommunications Act of 1996 211

10.6 Summary 212

A **Glossary**   **215**

B **Common Telecommunications Acronyms**   **219**

Biography   **227**

List of Figures

1.1   This is "basic?!?"   3
1.2   Data transaction from internet user to company server.   11
1.3   TCP/IP protocol implementation.   13
1.4   Comparison of OSI and TCP/IP protocol models. Notice that the "network" got smarter? Rather than just "dumb" layer 1 transport, the "network" today does lots of other things!   14
1.5   TCP/IP Internetworking.   16
1.6   SDLC private line.   17
1.7   PBX to PBX.   18
1.8   Token Ring user to local Ethernet user.   19
1.9   Local Ethernet to remote terminal.   20
1.10 Local Ethernet user to remote Ethernet user via Frame Relay.   21
1.11 Voice over Frame Relay.   22
1.12 SONET and ATM.   23

2.1   The LAN environment.   26
2.2   Basics of bridging.   28
2.3   LAN throughput.   31
2.4   Fast Ethernet. A typical 10 Mbps LAN will probably find that as traffic increases, the first segments to become congested are the links between LAN switches or bridges, and other "high traffic" paths such as those connecting to major services.   35
2.5   How does Fast Ethernet work?   36
2.6   ISO-Ethernet. Definitely worth a look! Provides 10 Mbps "P" Channel—plus—96 ISDN B Channels Integrated isochronous services for a LAN environment? Perhaps, but can ATM be far behind?   37
2.7   Isochronous/Nonisochronous.   38
2.8   FDDI—A "legacy" LAN technology? Sure looks that way. Standards have been around since 1989. Widely supported, but gradually losing ground to other technologies.   40
2.9   Circuit switching versus packet switching.   42
2.10 Packet versus cell switching.   43
2.11 Effect of ATM technology on an engineering application   44

2.12 Overview of network architectures. 45

2.13 Using LANs for a seamless solution. 49

2.14 Seamless solution—the 100,000 foot technical view of a hypothetical
      manufacturing company. 50

3.1   The WAN environment. 57

3.2   56 Kbps modem—"poor man's ISDN." 60

3.3   All digital network major points. 62

4.1   Everybody wants to connect!!! But first, what is the mission? 65

4.2   The mission is what best supports your customer. 66

4.3   Keys to success. 68

4.4   Vendor-managed relationship. 70

4.5   The management presentation is important. 71

4.6   Management needs the right information to make a decision. 72

5.1   Traditional T1 service (circa 1985). 83

5.2   Traditional LAN interconnection bottleneck. 84

5.3   T1 frame. 88

5.4   T1 framing overview. 88

5.5   Bit substitution. 91

5.6   DS1/ESF framing format. Extended super frame makes use of the framing
      bit by changing its meaning over time. The 8 Kbps that used to be wasted
      only on framing now provides two other useful features—CRC error check-
      ing and a data link channel. 92

5.7   M13 is *really* M1-2-3. 96

5.8   Overhead byte locations in an STS-1 frame. *Source:* ANSI. 100

5.9   SONET components. 101

5.10 What about the rest of the world? SONET does not hit every worldwide
      rate, but close enough. 102

6.1   SONET access ring 106

6.2   Connectivity, standards, and network management are the keys 112

6.3   Availability of advanced connectivity will change future network dynamics. 117

6.4   Graceful integration of emerging technology. 120

6.5   ATM cell structure. 122

6.6   Rewiring Frame Relay. 126

6.7   LANs? WANs? 128

6.8   ATM versus Frame Relay—who wins? 129

6.9   Isochronous/Nonisochronous. 130

7.1 Close collaboration no longer requires close proximity. 131
7.2 Managing nomadic users (understanding user requirements). 133
7.3 Help desk support versus type of applications. 134
7.4 The virtual office. 136
7.5 The virtual office, other considerations. 138
7.6 Jobs most easily adaptable to the home. 139
7.7 ISDN terminal adapter components. 150
7.8 Motorola's BitSurfer Pro. 152
7.9 Jetstream Communications' The Front Desk. 153
7.10 Siemens/Rolm's Officepoint Communication System. 154
7.11 ISDN can be like three lines in one! 155
7.12 ISDN example # 1. 156
7.13 ISDN example # 2. 157
7.14 ISDN example #3. 158

8.1 DSL to the home. 163
8.2 How fast is fast? 164
8.3 T1 versus HDSL configuration. 164
8.4 Three new profit opportunities for cable companies. 170
8.5 The $2 window, traditional resale. 171
8.6 The $5 window, new services for the SOHO market. 172
8.7 Direct satellite broadcast—Internet access. 176

9.1 Communicating business vision. 182
9.2 Graceful integration of emerging technology. Regulatory approval will be a key component. 185
9.3 The players today. 189
9.4 Recent changes. 190

10.1 Why "Telecom Reform" is not "reforming" anything—what can your company do? 200
10.2 Relative service contributions. 203
10.3 Residual ratemaking. 204
10.4 Who creates our national telecommunications policy? 205
10.5 Making it work. 207
10.6 Sample cost comparisons. 209
10.7 Strategy for realizing the vision. 211
10.8 Four options. 212

List of Tables

1.1   Terminology. 2

2.1   How fast can you run on a wire? 33
2.2   Gigabit Ethernet range problem. 41
2.3   Planning capacity for multimedia. 48

3.1   Kinds of dial-up modems. 57

5.1   U.S./International digital hierarchies. 83
5.2   Extended super frame overview (assignment of S bit in DS-1 signals). . . . 93
5.3   M13 frame. 98
5.4   C-Bit parity. Each C bit occurs 9,398+ times per second (9.4 Kbps). 98

Preface

Attention mainframe managers. Don't retire just yet. Do you think that all of those users who have defected to LAN platforms over the years are gone forever? Think again! The exodus by former mainframe users to distributed processing platforms has brought with it a whole host of new problems and complications in the operating environments they have created. In fact, the phenomena has created a whole new class of user, that is, the seat-of-the-pants LAN administrator. This new breed is business oriented, productive, technically astute, close to the business, and also—incidentally—about to tear his hair out by the roots.

Besides, when one pulls out a crystal ball and looks ten years into the future, one still sees a mainframe. It may, however, look like a huge server, and be controlled by personnel very different from those that occupied the Information Services "glass house" in days gone by.

The environment has changed, but have costs really gone down? Or have they just shifted from the IS organization to the business units? Has the shift to LANs been all it's cracked up to be, or are there other economies to be derived? Let's talk about that for a moment.

You see, the problem is that the local area network environment, once touted as everything from the key to greater productivity to a cure for warts, is much more difficult to manage in practice than on paper. In fact, the LAN manager is beginning to act suspiciously like the old mainframe jocks, much to the delight oftentimes of the latter. So what happened? Well, for one thing LAN managers are outgunned and undermanned. These platforms are people intensive and time consuming to operate. Availability, while decent by today's standards, often pales in comparison to the glass house. Ninety-eight percent uptime in the mainframe world was grounds for burning down the CIO's office, while in distributed processing, eighty percent or less may be tolerated in some businesses. The work force is also young. Great technologists, but oftentimes not as seasoned as one would like in core business issues. Instead they opt to be more enamored with the technology itself. Many times, in the eyes of executive management, we have torn down the glass house all right, only to replace it with an entire new work force of people dinking around with their PCs rather than making money for the company!

The new mission of tomorrow's technical services departments will be to "sell" high level network management services to Client-Server customers essentially "black boxing themselves to death." Many potential business units are having second thoughts about the

LAN business, as well as management of technology in general, including sentiments such as:

- Can we really afford to chase technology?

- Do we really *want* to chase technology?

- Wouldn't it be better to focus on our core business instead?

The extent to which you can address these sentiments will be a key component of how well your company will do in implementing tomorrow's systems. In other words, keep core business people focused on core business; keep technologists focused on technology; provide higher capacity network services than your user can provide on his own; provide better network management than your user can on his own. The payoff is a productive organization, effective standards for operation, easier help desk support, and lower cost. Once again, the hook in all of this will be providing and managing the wide area network.

This book will discuss how foresighted companies are proactively exploiting the opportunities attendant with internetworking by designing advanced networks and management systems, reclaiming a beleaguered user base, and in the process creating new efficiencies for their organization.

Leo A. Wrobel

1

First, the Basics

1.1 LANs, WANs, What Is the Difference?

LAN means Local Area Network, or basically "everything inside." WAN means Wide Area Network, or basically "everything outside" (the telephone company). There have been other manifestations of the name as well. Campus Area Network (CAN) is sometimes used, and the meaning is self-evident. Metropolitan Area Network or MAN is also sometimes used, sometimes in the context of a citywide LAN. (Wait, I thought LANs were inside!) Some of the more silly network acronyms include DAN (Desk Area Network) and PAN (Personal Area Network).

   Of course there is also POTS, which is not a network at all, but instead stands for "Plain Ordinary Telephone Service" and really is a term used in the industry. Therefore one can argue that we use pots and pans. This can get just about as silly as you care to make it. See Table 1.1. All kidding aside, however, we mostly use two terms: LAN and WAN.

   The other point that you have probably already deduced is that the acronyms in our business of networks will just about drive a new user crazy. Do not worry, we will try to acclimate you to the new language in the course of the book through repetition and immersion. By the time you are through the first few chapters, we will ask you if you can run B8ZS and ESF on an xDSL T1 circuit with CCIS7, and you know what? You may not know for sure but you will be able to offer an educated guess! If you get really stuck in the meantime, there is a glossary and list of common acronyms in the back of this book for your review. Use them as needed as you read.

1.2 So This Is Basic?

For a chapter that is supposedly dealing with the basics, where do we get off introducing a figure like the one on page 3? If you are new to telecommunications it probably does not look that basic to you! And you are probably right.

   Imagine, however, that you really wanted to learn Spanish. The best way to do it quickly would be to live with a family in Mexico City for three months, and completely immerse yourself in the language. That is kind of what we are trying to do here. By repeating the acronyms, and demonstrating a particular component's role in the big picture, we hope to help you remember these things by framing them in a context.

LAN	Local Area Network
WAN	Wide Area Network
MAN	Metropolitan Area Network
CAN	Campus Area Network
FAN	Foreign Area Network
DAN	Desk Area Network
PAN	Personal Area Network
POTS	Plain Old Telephone Service
POTS and PANS	Kitchen Implements

Table 1.1: Terminology. All kidding aside, all of these terms have been used at one time or another. Realistically speaking, we are most concerned with LAN and WAN.

In the same way that someone immersing themselves in a language learns quickly, having a framework in which to reference the words, rather than just a book from which to memorize them, will help you to learn quickly. That is also why some of the more important information repeats, albeit under a different context each time. It is the best way I have found to bring an overwhelmed technical service manager up to speed quickly. So with this in mind, let's evaluate the major components in this framework, in the context of a complex network diagram.

1.2.1 What Do Mainframes Do? (1)[1]

Mainframes do a couple of things really well. They are the cheapest way to deploy lots of computer (CPU) cycles to a large number of users. Mainframes are also good for applications that encompass repetitious tasks which are easily automated. Examples include airline ticket sales, rental car sales, and just about any kind of business where customers call in or walk in, and where an agent fills in fields in a form to sell a service. Insurance policies are another good example. Everyone fills in the same information: name, date of birth, address, and so forth.

If your business is characterized by any of these processes, mainframes are probably the ticket. We will discuss more about mainframes, and their role in future networks later in this book.

1.2.2 What Is a Front-End Processor? (2)

You can imagine that a mainframe supporting hundreds of telemarketers, for example, is probably a pretty busy box. Therefore, mainframe manufacturers invented a "traffic cop" to

---

[1]In Chapter 1, all circled numbers appearing beside section headings refer to the circled numbers in Figure 1.1.

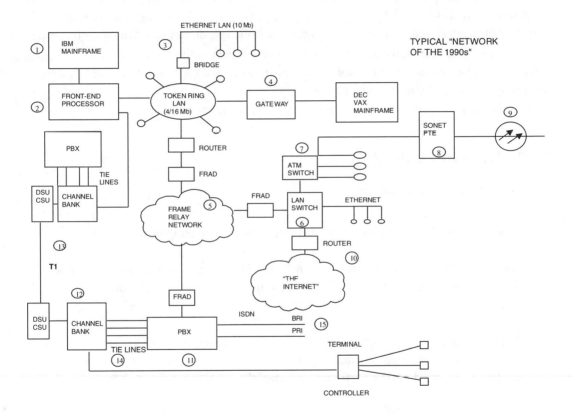

Figure 1.1: This is "basic?!?"

direct all the messages from the users into and out of the mainframe to avoid bogging the mainframe down. The front-end processor is therefore just what the name implies, it is a gatekeeper to the mainframe.

Now with this in mind, you may already know that there are other boxes that perform a similar function in today's networks. Routers and LAN switches come to mind. And indeed, many of these devices are overtaking and usurping the front end due to their flexibility and ease of integration into distributed LAN networks. For this reason, the front end might just be an endangered species. When you look into your crystal ball, ten years from now you will still see a mainframe. They are just too good at what they do and just too inexpensive in a large organization to dispense with. Front ends, however, do not seem to be on the radar.

By the way, a common name for both of these boxes is the *legacy* network. Legacy, as the name implies, means the "holdover" technology from a bygone era, and is loosely interpreted to mean the mainframe environment.

So what is replacing the "legacy" network? It is the LAN network.

1.2.3   *What Are LANS?* ③

LANs are characterized by the following:

- They are *local*, meaning within a building or at most a campus environment.

- You, not the telephone company, own them. That means the bits are "free" once the initial investment has been made. It also means that in case of trouble you are on your own.

- LANs usually run on copper cable media, like twisted pair or coaxial cable. Older technologies used coaxial cable, while most new ones use unshielded twisted pair (UTP) cable rated for the particular speed of the LAN.

The two most prevalent LAN technologies out there today are Ethernet and Token Ring. Ethernet was designed to run at 10 Mbps, but has been all but replaced in large organizations by Fast Ethernet, which operates at 100 Mbps. This has chiefly been the result of a major drop in the price of Fast Ethernet cards and greater bandwidth (capacity) demands of the user.

Token Ring was originally embraced by IBM. It too is being replaced in most organizations by Fast Ethernet. Token Ring operates at 4 or 16 Mbps. It has been commonly held that if the decision on LAN technology was made while the mainframe bigots were still in charge, the company went with Token Ring. In later years, the swing was to Ethernet.

Both of these technologies will be discussed in greater detail in Chapter 2.

1.2.4 *How Do We Connect LANS?* ④

In much the same way that a front-end processor is the gatekeeper to the legacy environment, a bridge, router, or gateway is the gatekeeper to the LAN environment. Bridges basically isolate and filter different LANS from one another. This way if one LAN segment gets really busy, it does not slow down the whole organization. On the other hand, if the two segments need to communicate with one another, the bridge will forward a message to the other segment. That is why people commonly describe the function of a bridge as two things: filter and forward.

Routers and gateways act similarly but have a few more brains that bridges. For example, bridges are not particularly good at dealing with ambiguity. If more than one path exists to a user (common since networks are often engineered to provide multiple paths for fault protection), a bridge is not particularly well suited for selecting the best one. A router on the other hand is much more capable in this area.

1.2.5 *What Is Frame Relay?* ⑤

Frame Relay is one of the hottest technologies available today. In Figure 1.1 we have it illustrated as a "cloud." I hate clouds. My first question to a vendor is what kind of network is inside the cloud, and in later chapters we will look into exactly that.

So why is Frame Relay so hot? Imagine for a moment that I am a competitive *postal* provider and offer you a deal. I will deliver a letter anywhere in the U.S. for five cents. Sound interesting? Good, I will go on.

In order to get this rate, you have to follow some rules. First, you will have to commit to some minimum volume of letters you will let me deliver. Second, you will need to use *my* envelopes. Don't worry, I'll provide you with the envelopes. Third, you can put no more than one sheet of paper in each envelope. If you want to send a three-page letter, you must use three envelopes. Even so, that is still only 15 cents, less than half of what the U.S. Post Office charges. For 100-page documents from the legal department, you can always drop them in Priority Mail using your old postal carrier.

In keeping with this analogy, consider the following. If you mail five letters to Dallas on the way to work in the morning, then five more at lunchtime, then five more at 5:00 p.m. on the way home:

A. Will all 15 arrive in Dallas at the same time?

B. Will all 15 get to Dallas in the same way?

C. Will all 15 even get to Dallas on the same *day*?

D. Is there any guarantee that all 15 will even get to Dallas at all?

The answers to A, B, C, and D above are all "no." And the same holds true for Frame Relay.

First off, 15 data packets sent on a Frame Relay network to Dallas will not necessarily get to Dallas at the same time. That is because not all 15 will take the same path through the network. The carrier decides on the path based on network conditions at the instant you send the data. Just like your letters may take a train, aircraft, or truck depending on the time you send them, your data similarly will take different paths.

There is also no guarantee that the data will get there at all. If the Frame Relay network gets too busy, a few packets might be thrown overboard. But when you think about it, even with a dedicated circuit there is no guarantee that your data will get to the far end. That is why there are higher level protocols like TCP/IP or SDLC which check for lost data or packets delivered in the wrong order. That means Frame Relay works great for data, and it is cheap. But what about voice or video?

Voice and video require a continuous bit stream with the timing relationship of the bits maintained. Another word for this is *isochronous*.

For example, if I am watching the Dallas Cowboys on TV here in Dallas, and Emmitt Smith rushes for a touchdown, I get to see it about half a second earlier than someone in New York who would be receiving their view of the game via network television satellite. We see the same play however, and even though the New York viewer sees the play delayed by half a second (because of the propagation time to the satellite) he or she sees the same thing I do.

Imagine however that the video image was sent over a Frame Relay network, where there were no guarantees that the bits would arrive in order. I could just hear John Madden now:

> He's at the 40!
> The 30!
> The 35!
> The 20!
> The 5!
> The 10!

Okay, so I made a lame joke. However, this does illustrate the point and why a user contemplating voice or video on a Frame Relay network, or other *non*isochronous means of transport could have problems. We will discuss this technology in greater detail later in the book.

1.2.6 *What Do LAN Switches Do?* ⑥

For lack of better words, LAN switches are souped up bridges. Not only do they filter and forward, but they have advanced network management capabilities which allow the technologist to keep an eye on the network through proactive maintenance. Moreover, LAN switches are *fast*, and for that reason they often replace routers. Lastly, today's LAN switches allow myriad possibilities for mixing and matching different equipment, components, and environments for truly seamless solutions. These are discussed later in the book.

1.2.7 What Is ATM? ⑦

What is a user to do if he desires the economies offered by Frame Relay, combined with isochronous capabilities for voice and video? What if a user wants to eliminate the distinction entirely between his LAN and WAN environments? What if a user needs to move multiple gigabyte diagrams, such as computer-aided designs or X-ray images? What if users need a single, integrated network access for voice, data, video, image, multimedia, and so on, with virtually limitless speed and flexibility? These describe the characteristics of the ATM user. If comedian Tim Allen was allowed to "rewire" Frame Relay, he might come up with ATM.

Why do LANs and WANs have different rules? When ATM becomes prevalent, the signal will stay the same from a workstation in New York to a workstation in Dallas and the distinction between LANs and WANs will be virtually gone. We will describe how this will happen in later chapters.

1.2.8 What Is SONET? ⑧

SONET will replace T1 and T3 as the backbone transmission media in the U.S.

Since T1 and T3 have been around for almost 40 years, it sounds like a safe bet that by the time there is a replacement for SONET, we will all be wandering around shopping malls looking for the perfect frozen yogurt. And how often do we get a planning window that long? That is the thing about transport technologies. They have a nice long life.

SONET stands for Synchronous Optical NETwork, and was designed as a transport technology for fiber optics. SONET is far smarter than T1. T1 was originally designed as a means of combining 24 channels between Bell central offices onto four wires instead of 96. T1 was designed for *voice*.

T1 is woefully dumb. In fact, if you put a T1 and a cinderblock into an intelligence contest, it would be a tie.

SONET on the other hand is *smart*. SONET is clean and error free. SONET has the capability to repair itself if the fiber is cut through a feature called fault protection switching. SONET has nonintrusive maintenance features—meaning you do not always have to take it down to test if it has trouble. And SONET is designed for glass—fiber optics—the mainline telecommunications medium in place today.

Your users will be willing to pay for the features in SONET, and the technology is more available today than ever. We will discuss the features of this exciting and long-lived technology, as well as how to get it into your organization in later chapters.

1.2.9   What Is Fiber? ⑨

Fiber optics is the medium used for virtually all high capacity communications services. By high capacity, we mean essentially anything above the T3 rate of 44.736 Mbps. Fiber is also used in the LAN environment as a high speed connection between LAN switches and other components. Fiber can also be used to connect mainframes together at the channel level in the legacy environment. There are two kinds of fiber:

1. Single Mode. This is what the phone companies use.

2. Multimode. This is probably what you have in your building.

We will discuss their differences in later chapters.

1.2.10   What Is the Internet? ⑩

Imagine you are the U.S. Government in the 1960s and 1970s, planning for the prospect of incoming Soviet ICBMs. Intuitively, the war planners of the time knew that 20 minutes after a launch, virtually every telecommunications hub in the U.S. would probably be in the upper atmosphere. This presented an obvious problem.

Since any potential opponent would probably hit telecom hubs first (thereby taking out the eyes and ears of their opponent) some survivability had to be planned for. The first thing the government did was build some big telephone centers in relatively outlying locations. For example, consider Ennis, Texas, population 14,000, about 30 miles south of Dallas. A nice little town, good Czech sausage, and one of the biggest AT&T offices I've laid eyes on. I'm not sure if it was part of the 1960's effort to harden the public network, but it sure fits the bill: uncharacteristically big for the town, no windows, and 30 miles from ground zero.

Even with precautions like these however, it would be difficult for data communications to take place using the traditional "master-slave, host-terminal" environment. It would be logical to assume the host would be in a big city. That means that even with the path in place, data communications would still be out of the question.

Just what if, for example, each node in the network was just a little bit smart? Distributed intelligence with nobody in charge. (Kind of like my office.) The idea behind the Internet was that data could be put in packets, with a header address to where it was destined. This packet is called a *datagram*.

After a first strike, some parts of the network might still be intact. Using a distributed intelligence topology like this one, messages might eventually get through. This was the beginning of the Internet we know today.

Routers are just a little bit smart. They are better at dealing with ambiguity than bridges. For example, they can select an "alternate" path when the usual one is out based on internal lists called forwarding tables. Routers, like everything else, require special software called

TCP/IP. Routers do not care what the data is. They just route datagrams to the next router available.

When you access a Web server today, this "bucket brigade" system of distributed intelligence routes datagrams to and from their destination. Only today the government is out of the picture, the Web is commercially supported now. And nobody owns it. Since the Internet is in fact a collection of servers, "distributed intelligence with nobody in charge," it is difficult to control. This will be a policy decision taken up later; for example, can a company put mission critical data on a network that nobody controls, and where there are no quality of service guarantees? On the other hand, the World Wide Web has more users today after just 10 years than the telephone network had after 50. It is clearly positioned to be revolutionary. We will delve into this topic in greater detail later as well.

1.2.11   What Is a PBX? ⑪

Much of what we have covered so far has been data oriented. PBXs are for voice. PBX stands for Private Branch Exchange. It is a private telephone switch owned and maintained by your organization, on its own premises. It is also probably a misstatement to say that PBXs are only for voice. Some of them also do data, video, and so on. This is not surprising however. In today's networks, voice is really data. So is data. Any questions? Relax, we will cover the distinctions in this (true) statement later in the book.

1.2.12   What Is a Channel Bank? ⑫

Channel banks take a T1 (described below) and break it into 24 64-Kbps voice channels. This is what T1s were originally designed for. A T1 was a cost effective means of taking 24 interswitch trunks between Bell or AT&T central offices, and combining them onto a four wire copper circuit. Since the technology worked so well, and since it has been out there for so many years, it has found its way to the end user as well. Today virtually all end users utilize T1 access lines to combine dial tone lines, trunk lines, and other voice and data services.

1.2.13   What Is a T1? ⑬

If you take 24 64-Kbps channels, and multiply the numbers together you get 1.536 Mbps. Add 8 Kbps more that the phone company uses to tell them all where to line up and you get 1.544 Mbps. That is the T1 speed. In fact, you can think of a T1 as a 1.544 Mbps *data circuit*. Twenty-four circuits share this circuit, and each one gets exactly 1/24 of the capacity. There is a T1 section in this book as well. About halfway through it you will probably want to shoot me, or shoot yourself, but the level of detail is important to illustrate what is unquestionably the most widely deployed transport technology today. Moreover, it will provide the basis for comparison with what has changed with SONET, the technology which will replace T1, and how we can best exploit this change.

1.2.14   What Are Private Lines? ⑭

T1s will carry more than switched voice traffic. They also carry "private lines," otherwise known as leased lines, custom circuits, or by other names.

A 64 Kbps voice channel can be used for other things than voice. One would be a digital data circuit operating at 56 Kbps or 64 Kbps. Essentially, creating one of these is no more difficult than plugging the appropriate channel card into the T1 channel bank.

Rather than expecting off hook and on hook information from a PBX, the digital data card tells the T1 to expect data bits in that particular time slot on the T1.

On the other hand, the phone company or user might choose to plug a four wire analog circuit card into the channel bank. The card is still designed to carry data, but is looking for an analog signal now instead of a stream of digital bits. This would require a modem.

There are over 200 different kinds of private line circuits. These range from "hoot and holler" lines to digital data circuits.

A hoot and holler line is pretty simple; in fact, it is just a pipe. It does not even ring. The user just picks it up and yells, hence the name. "I want to trade 100 shares of AT&T at 52!" or "I need a fender for a 1964 Chevy!" Junkyards and stock brokers often use this kind of circuit.

1.2.15   What Is ISDN? ⑮

ISDN is designed to replace POTS (remember, that is plain ordinary telephone service) by bringing digital to the doorstep of the small office/home office (SOHO) user. Remember, that 64 Kbps channel we spoke of? Well, it can do more than just carry a voice. We have already shown that a channel can carry data at 64 Kbps as well. ISDN takes this concept a step further.

The copper cable which serves the typical small office or residence is really capable of carrying more than just a single voice call. ISDN BRI (Basic Rate Interface) takes advantage of this by providing *two* 64 Kbps channels on the same two wires which previously carried POTS. In fact, it looks and acts like a miniature two channel T1. The common name is 2B+D. That means two bearer channels with one delta channel. The "B" channels carry the actual information. The "D" channel can be thought of as the "supervisor" that instructs the two "worker Bs" what to do, how, and when.

ISDN also comes in a large size. Primary Rate Interface or PRI can be termed "T1 Deluxe." It looks, acts, and quacks much like a T1, but has some additional features. It has 23 channels that carry either voice or data traffic, plus a 24th channel (another supervisor) that controls the whole show and tells the rest what they are and what to do. The common phrase is 23B+D, or 23 bearer channels with one delta channel.

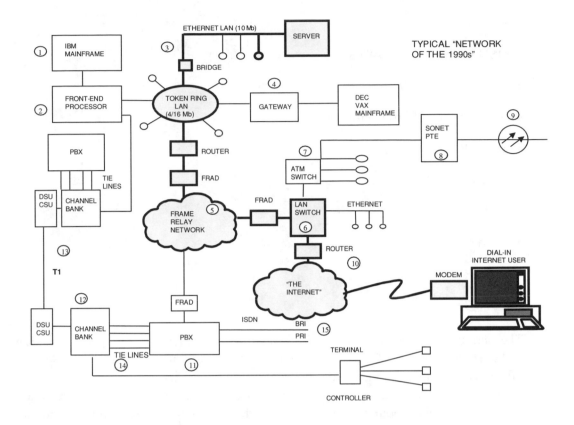

Figure 1.2: Data transaction from internet user to company server.

For your convenience in remembering which is which, just think BRI and PRI. B is for baby, P is for Papa. Both technologies are enjoying broad acceptance. For now however, PRI is used primarily for PBX access just like a smart T1. BRI, however, is surging in use as a means of surfing the Internet at 64 or 128 Kbps.

1.3 How Does Data Flow Through This Network?

Okay, now that we have all the players formally introduced, how does data actually flow through this network? For that to happen, we need to have a set of rules, otherwise known as a protocol. The problem is, there are many protocols. Figure 1.2 shows the devices which utilize the protocols shown in the following list.

- Mainframe—SNA
- Legacy WAN—SDLC

- PBX—PCM

- DSU/CSU—AMI or B8ZS

- Channel Bank—D4 or ESF

- Frame Relay—Carrier Proprietary

- Token Ring

- Ethernet

- Fast Ethernet

- ATM

- TCP/IP

- SONET

The permutations are endless. For illustrative purposes, let's trace a data transaction using the TCP/IP protocol from a homebound user dialing into an ISP (Internet Service Provider) to a server on Ethernet LAN shown as ③ in Figure 1.2.

First it is important to understand that the protocol will change many times in transmission. One way to think about this is to picture the data being put into an envelope, and that envelope either delivered, or stuffed into yet another envelope that allows another system of delivery (see Figure 1.3, TCP/IP protocol implementation).

The story begins with the Internet user, who generates the data. This has no address or envelope, it is just data. This data drops into the first envelope called a TCP segment. TCP puts an address on it, and hands it to IP.

TCP/IP can be thought of as the Abbot and Costello of telecommunications. If you think back to the old movies, Costello was the screw up. Costello could lose things. But he always had his buddy Abbot to get him out of jams. That is kind of what TCP/IP does. IP is the screwball. He can lose things or deliver things out of sequence. TCP, on the other hand, keeps tabs on IP to assure data is delivered, delivered in the correct sequence, and delivered without errors.

IP takes the TCP segment and drops it into an IP datagram. The datagram is the most elementary routable data packet. IP neither knows nor cares what is in the packet. It just routes datagrams.

Since the user is on a dial-up modem, he is using a PPP (point to point protocol) to keep things in order on the dial-up line. Therefore, the datagram drops into the PPP frame, and is transported via the dial-up connection to the Internet Service Provider (ISP).

With me so far?

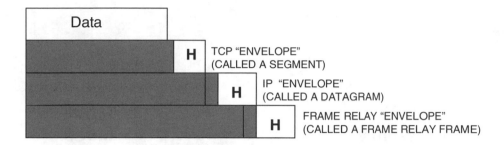

Figure 1.3: TCP/IP protocol implementation.

Once at the ISP, the equipment there opens the PPP frame and finds the datagram. The datagram is routed through the Internet via any number of routers. Based on information in the routing tables, the routers figure out an optimal path given network conditions at that instant in time. They route to an IP address specified in the header of the datagram which corresponds with a port on the LAN switch ((6) in Figure 1.2).

The LAN switch has a router card in it that can move the datagram based on information in its routing tables. This tells the LAN switch that the most optimal path is via the Frame Relay network.

In order to move the datagram over the Frame Relay network, it will be necessary to generate a new envelope. The FRAD (Frame Relay Access Device) opens a new Frame Relay frame, and inserts the datagram in the Frame Relay envelope. This envelope is routed through the public Frame Relay network based on network conditions at that instant to a DLCI (Data link connection identifier) at the far end which corresponds to a FRAD at the other company location.

The FRAD speaks Frame Relay out one side, and IP out the other. It opens the Frame Relay envelope and finds a datagram. This is much to the delight of the IP side of FRAD, which says, "Hey this is a *datagram*, and my reason for existence! I *route* these!" and promptly hands it to the next router upstream.

The next router speaks IP out one side, but Token Ring out the other. You see, to get to our destination we now need to open a Token Ring frame. The router generates a Token Ring envelope, and sends it over the Token Ring network. At the other end is a bridge.

The bridge speaks Token Ring out one side, and Ethernet out the other. It opens the Token Ring frame, and generates an Ethernet frame. It pulls out the datagram, places it into the Ethernet envelope, and sends it to the server.

The network interface card (NIC) in the server matches the MAC address (Ethernet address) on the Ethernet frame. The NIC opens the Ethernet envelope and out pops a datagram. IP software hands the datagram to TCP. TCP opens the IP datagram, assures the

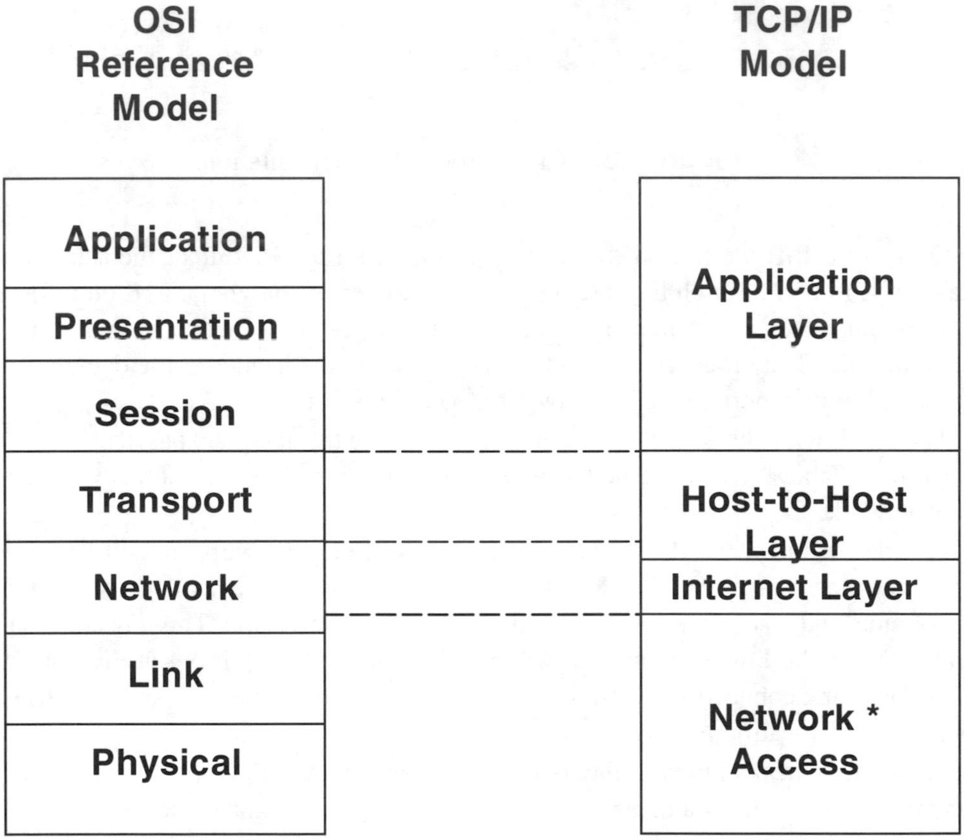

OSI Reference Model		TCP/IP Model
Application		Application Layer
Presentation		
Session		
Transport	- - - -	Host-to-Host Layer
Network	- - - -	Internet Layer
Link		Network * Access
Physical		

Figure 1.4: Comparison of OSI and TCP/IP protocol models. Notice that the "network" got smarter? Rather than just "dumb" layer 1 transport, the "network" today does lots of other things!

data is error free, and hands the segment to the application. The application in turn pulls out the original data, which in this case might be a request to access a Web page on the server by the original user.

So as you can see, in addition to lots of hardware, software interactions take on a significant role in the transfer of data. Protocols change innumerable times. It is no wonder why networks become so complex and why users often "black box" themselves to death! Figure 1.4 might help put things into perspective on the software side.

The OSI reference model is shown on the left. T1 lives in layer 1—the physical layer. Ethernet and Token Ring live in layer 2. TCP/IP lives in layers 3 and 4. The application lives in layer 7. The TCP/IP model on the right is simpler and shows only four layers, the application layer, TCP layer, IP layer, and network access layer.

Notice that the network layer got bigger? What used to be just a dumb layer 1 network layer in the OSI model now goes to layer 2 1/2. This is because networks have become smarter. Frame Relay (Figure 1.5), for example, does routing, switching, segmenting, error control, and other higher level functions that used to be I.S. rather than telecommunications functions. We will discuss the implications of these smarter networks later as well.

The "legacy" version of data transfer is shown in Figure 1.6. Mainframes and front ends use SDLC, and the applicable components are highlighted for your review. Figure 1.6 also shows the "envelope in an envelope" example in a comparison of IP and the U.S. Post Office.

1.4   PBX to PBX

Voice communications between PBXs are handled differently as shown in Figure 1.7. The backbone link between the two is a T1. The T1 terminates in a DSU/CSU that converts the digital signal from the T1 to one that can be understood by the PBX. The DSU/CSU is an integral component of the channel bank, or can even be internal to the PBXs themselves. If a PBX has a T1 card, a channel bank is not necessary, nor is a DSU/CSU. This is shown in Figure 1.7 as the "traditional" way to break a single T1 into 24 individual channels—in this case, *tie lines* between PBXs. Tie lines are used to connect two PBXs in neighboring buildings, or even different cities. One benefit of using them is the ability to provide four or five digit dialing transparently and simultaneously in both locations.

1.5   Token Ring User to Local Ethernet User

Figure 1.8 is highlighted to show a local Token Ring user connecting with a local Ethernet user via a bridge. Bridges are now smart enough to not only do the filtering and forwarding but to also do the format translation (Token Ring to Ethernet).

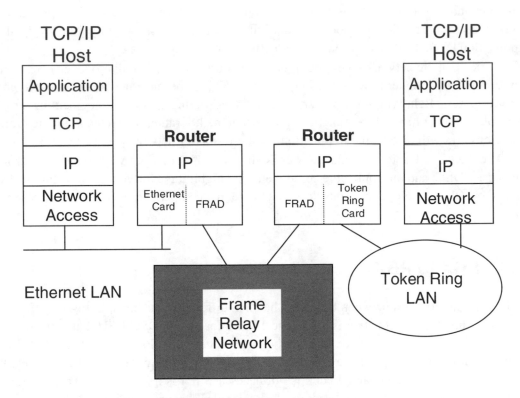

Figure 1.5: TCP/IP Internetworking.

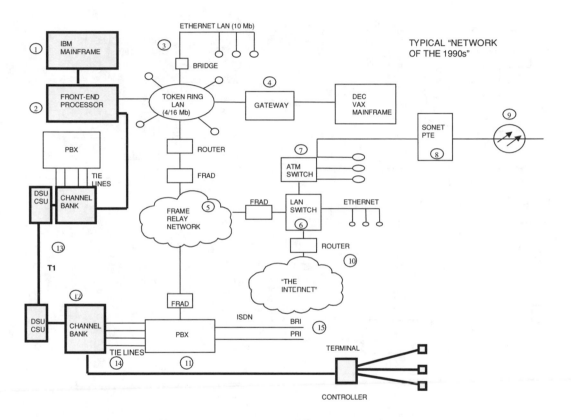

Figure 1.6: SDLC private line.

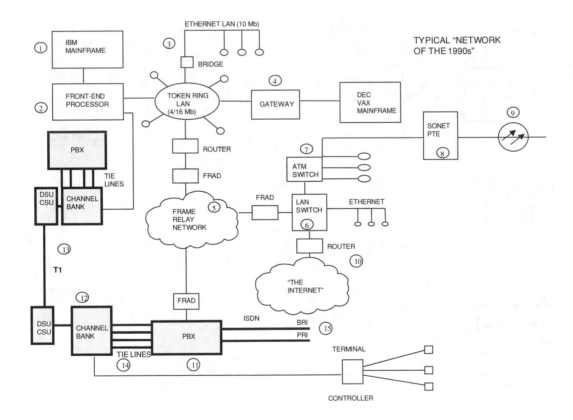

Figure 1.7: PBX to PBX.

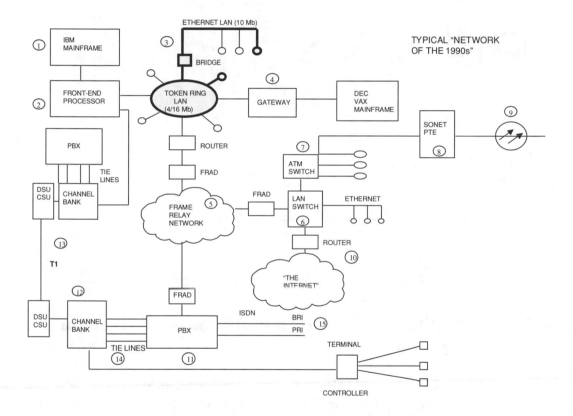

Figure 1.8: Token Ring user to local Ethernet user.

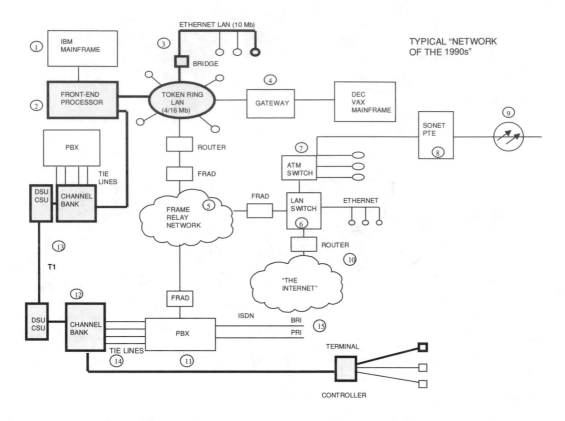

Figure 1.9: Local Ethernet to remote terminal.

1.6   Local Ethernet User to Remote Terminal User via Private Line

Imagine that the Ethernet user needed to communicate with a distant SDLC terminal device. Several format translations must take place, but through each of them (Ethernet→Token Ring→SDLC) the data stays the same as shown in Figure 1.9. Only the envelopes change.

1.7   Local Ethernet User to Remote Ethernet User via Frame Relay

Frame Relay is a wonderful technology for connecting LAN users together. As we stated earlier, Frame Relay is not *isochronous*, but that is okay. Neither are most LANs. Additionally, if traffic is discarded due to network conditions, the intelligent higher level data protocols simply retransmit. This is illustrated in Figure 1.10.

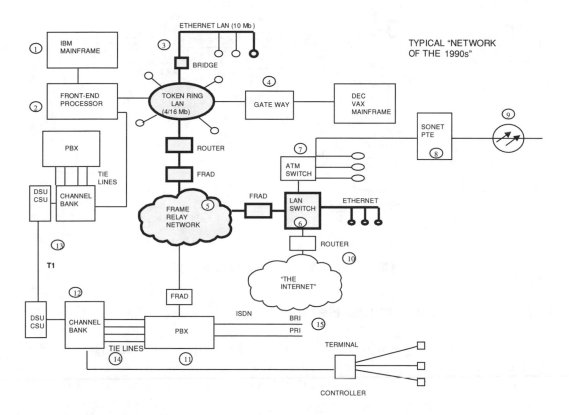

Figure 1.10: Local Ethernet user to remote Ethernet user via Frame Relay.

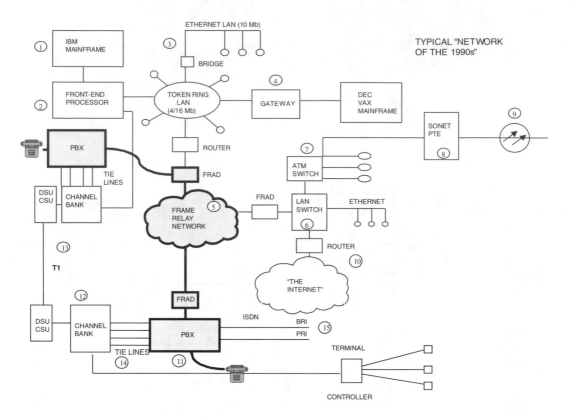

Figure 1.11: Voice over Frame Relay.

1.8   Voice Over Frame Relay

Unlike data, voice poses problems for Frame Relay. In the case of "discard eligible" traffic, which part of my voice is not important? Voice is less forgiving, and in fact, is the *premium* kind of traffic on most kinds of Frame or Cell Relay. For those of us used to cheap voice and expensive data, this represents a major change. This is illustrated in Figure 1.11.

1.9   SONET and ATM

How can one derive the benefits of frame- or cell-based networks but still have *isochronous* capability? For these kinds of applications, ATM will be the ticket. And ATM will ride on SONET. See Figure 1.12.

For the moment, think of SONET as railroad tracks—large fixed investment with carrying capacity dictated by the number of trains run in a given period. ATM cells are the trains. Even though the capacity of the tracks (SONET) stays the same, the carrying capacity goes

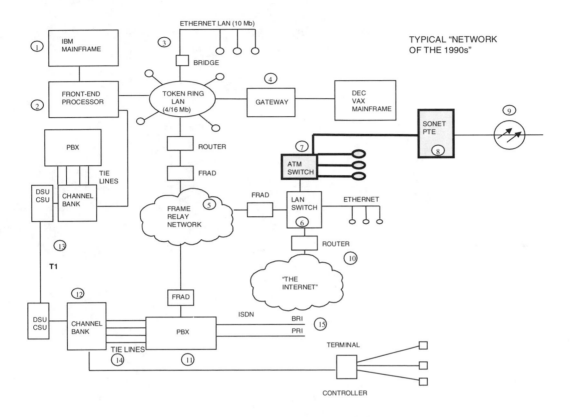

ETHERNET LAN (10 Mb)

TYPICAL "NETWORK
OF THE 1990s"

Figure 1.12: SONET and ATM.

up as more ATM cells are sent. ATM therefore provides the perfect way to share SONET bandwidth. When some users are not actively using the tracks, others can run trains—on a second by second basis!

Now that the basics are out of the way, let's take a closer look at the *Local Area Network* in Chapter 2.

2
LAN Basics

2.1 What Is a LAN?

Although this book is primarily a Wide Area Network (WAN) book, the lines between the LAN and WAN become narrower every day. Therefore, in order to effect truly seamless integration of both network environments, at least a rudimentary knowledge of the LAN environment is necessary for a network engineer or planner.

So what are the characteristics of a Local Area Network?

Local Means "Limited Geographic Coverage:" LANs are designed to support transmission ranges under three miles and are intended for use in a single building or campus.

Customer Owned Cable—Bits Are Free: Unlike wide area data networks where the user rents a transmission channel from a common carrier, in a LAN configuration the user buys and installs the cable used to connect the various devices. In this way, there is no ongoing cost for the communications channel.

High Transmission Rates/Few Errors: LAN stations transmit at bit rates over 1 Mbps and operate virtually error free. In a LAN, that transmission capacity may be dedicated to one user or shared by a group of users.

Designed for Data Applications: Most traditional LANs like Ethernet and Token Ring are not designed to support voice or video transmissions. Some of the newer LAN technologies like ISO-Ethernet and ATM will support voice, data, and video.

See Figure 2.1 for a typical layout of a LAN environment.

2.2 LAN Operations "Refresher"

If you consider the workgroup illustrated in Figure 2.1 you get an understanding of what LAN administrators are constantly doing to maintain a reasonable degree of response time and performance on the LAN. Consider Figure 2.2. We illustrate a simplistic diagram of an Ethernet segment with 10 users on it, shown as users A–J, respectively. In the center section of Figure 2.2, user J begins to consume an inordinate amount of resources on the LAN due to large file transfers or other heavy activity.

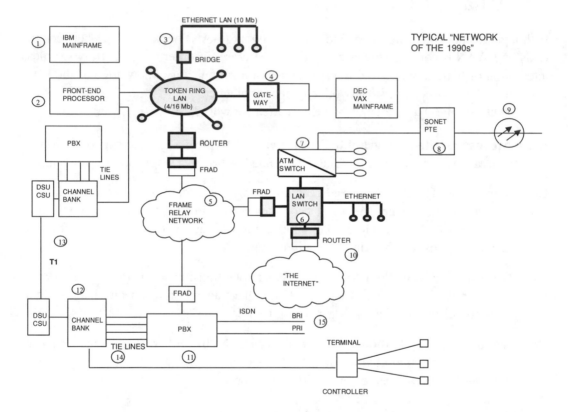

Figure 2.1: The LAN environment.

If the consumption of resources is great enough the other users notice a notable drop in response time and network performance and begin screaming.

Since it is impossible to create more capacity in an Ethernet segment we must instead opt for *segmenting* the network, or putting users on smaller pieces of the LAN so that resources are divided as many times as necessary. This is illustrated in the bottom section of Figure 2.2.

The easiest way to do this would be to cut the network in half between user F and user G as illustrated. The problem with this becomes apparent, however, in this case. Users A–F cannot access the server or cannot access any of the other users, G–J. So instead, a device is used between users E and F to *segment* the network, and provide a *filtering and forwarding* function to *forward* only messages destined for the other LAN segment and to *filter* any messages designed to access the other users in the immediate workgroup. This device is called a *bridge*.

The basic mission of a bridge is to segment LAN networks by providing a filtering and forwarding function, thereby improving performance of the network as a whole.

Bridges are not terribly intelligent devices, but they are very fast and quite versatile. They require little in the way of maintenance, and can usually configure themselves dynamically. A bridge will build a forwarding table in memory, identifying a hacker of which it has knowledge. Looking back at Figure 2.2, if A wants to send a message to D, users F–J have no need to be involved in this transaction. Therefore the message is filtered to users F–J by the bridge. If user A wants to access a file on the server however, this data must transverse the second LAN segment. Therefore the bridge forwards the transaction. The net result is a reduction in overall capacity on each LAN segment by filtering unnecessary transmissions.

Bridges move traffic around the LAN based on a 48 bit MAC address, which is hard coded into each network adapter card on the bridge port. Therefore there is no real addressing as in TCP/IP and other higher level protocols as data is moved around at the physical level. This makes bridges fast as opposed to routers. In fact, new and improved bridges are coming out now, otherwise called *LAN switches*.

One casual way to refer to a LAN switch is nothing more than a "souped up bridge," or a bridge that provides more in the way of segmenting functionality and versatility than a common bridge. Indeed LAN switches are replacing routers in many cases because of their ease of maintenance, their high capacity, and their speed.

Before we move on—*what is a LAN*? Basic characteristics of LANs, or local area networks include:

2.2.1 *Limited Geographic Coverage*

LANs are designed to support transmission ranges which are limited in serving area such as buildings and campuses. Typical distances are measured in hundreds of yards as opposed

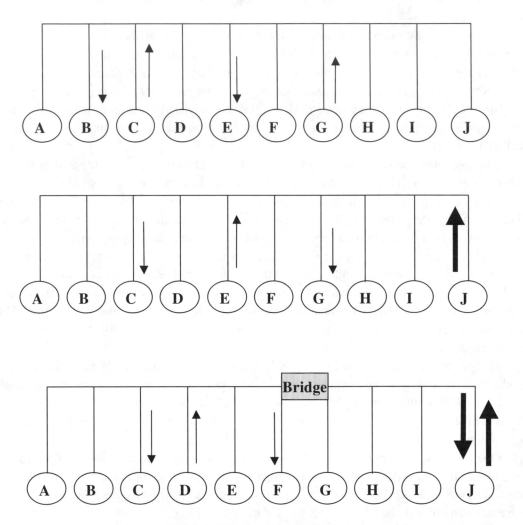

Figure 2.2: Basics of bridging.

to hundreds of miles. This differs from wide area networks where a vendor such as a local or interexchange telephone company takes responsibility for the transmission medium. In the wide area the telco is in control. In the local area you are in control, as described below.

2.2.2 LANs Are Customer Owned

The user buys and maintains the device to build the LAN. This is a two-edged sword. On one hand, there is no ongoing cost for the communication channel itself and the capacity is free. On the other hand, users are responsible for everything. LANs are characterized by very high transmission rates with very few errors. Errors in the LAN environment, if properly engineered, are very rare. In fact the medium can almost be considered error free. Some of the error figures we see for a properly engineered local area network are 10 per year, making it a very clean and high quality medium if properly installed. Even so, even minor variations in an installation can cause errors, such as running cabling over fluorescent lights, using the wrong types of jacks, or a generally sloppy installation.

2.2.3 LANs Are Designed for Data

Most traditional LAN technologies such as Ethernet or Token Ring are *isochronous* and do not lend themselves readily to voice and video transmissions for reasons we will discuss later. Therefore LANs are designed primarily for *data* or for types of traffic where a timing relationship of the data packets is not important.

Let me explain the term *isochronous* in a little more detail. Imagine you are doing a full motion video transfer of me, the author, at one of my speaking engagements. Typically a video link is *isochronous*, meaning that the bits leave the location while I am speaking and arrive at the location where you are viewing in sequence with each other and with the timing relationship maintained. Another example of *isochronous* transmission would be a voice telephone call. Once again, in order to move an intelligible voice from one end to the other, the words that leave the location from where I am speaking have to arrive at the location where you are listening in the same order. Imagine, in the case of the video conference and in the case of the voice call, if the video and the voice were broken up and all of these packets arrive at the far end out of sequence with each other. Imagine further that other users could jump in the middle and send their data at the same time thereby delaying packets in my transmission from one end to the other. This would mean the voice coming through on the other end would be unintelligible since the words would be out of sequence, phrases would be mixed, and everything would be out of sync. The video would be even funnier, as you watched me go from one end of the room instantly to the other end of the room, up on the ceiling, down on the floor, and every which way. I'm being a little precocious with this, but you get the idea why *isochronous* capability is important. Even 100 Mbps Fast Ethernet uses the same CSMA/CD access protocol and is prone to the same concerns regarding isochronous capabilities.

CSMA/CD stands for Carrier Sense Multiple Access with Collision Detection. Since we have not discussed that yet in the text, let's briefly describe CSMA/CD, and show why it is not isochronous.

Consider Figure 2.2. Let's assume that device A wants to send data to the server. Device A has a NIC or network interface card that "listens" to the LAN network. After ascertaining that no one else on the network is sending, it begins to send data to the server. Let's assume, however, that at the same instant the network interface card in C is also listening to the network. It hears that nothing is going on out there, so the network interface card in C begins transmitting at the same instant the network interface card in A begins transmitting. Since there is only one medium out there, the data from A and C *collide* with each other. This is what is known in a LAN environment as a *collision*.

Since the network interface cards continue listening while they are transmitting, they detect this collision and realize the data has been corrupted. At this point they stop and execute a predetermined routine internal to the card, which tells them how long to wait before sending again. Whichever device "wins," so to speak, with the lowest number (A or C), begins transmitting again. Hopefully the transmission gets through the second time. However, if a collision occurs again, the same process repeats. This describes the process of CSMA/CD in the simplest possible lay terms.

By considering this example, it is easy to see how performance problems manifest themselves in a LAN environment. Obviously, the more collisions, the lower the throughput. We can actually measure these types of things on special equipment called data-sniffers. Consider Figure 2.3 entitled LAN throughput.

How many users can occupy one Ethernet segment? The answer is probably five thousand or more assuming absolutely no one sends anything. Okay another joke, since this is obviously unrealistic. Let's take another approach.

The factor governing the number of users that can occupy an Ethernet segment deals principally with how often they send, and how big the files are. When LAN utilization approaches 25 percent, most LAN administrators consider this a "yellow light." At this point there is enough collision on the network that it becomes apparent to users that the network is busy and things are slowing down. As this continues, at about 40 percent utilization the network grinds to a virtual halt. This is because there are so many collisions, it becomes impossible to get meaningful throughput using a network. Therefore LAN administrators should keep a close eye on models such as the one illustrated in Figure 2.3. But, this is not always possible if the company has not invested in a management system up to the task. Therefore, rather than keeping such a close vigil, LAN administrators just wait for users to start screaming—then they segment the network further. Another point that can be inferred from Figure 2.3 is that even though Ethernet segments are rated at 10 Mbps, actual throughput in a busy LAN environment is probably 4 Mbps or less. Once again, even this speed is adequate for most applications, assuming the network is well engineered. However a "wild card" application, such as multimedia, throws everything into doubt. Lucky for us today,

UTILIZATION

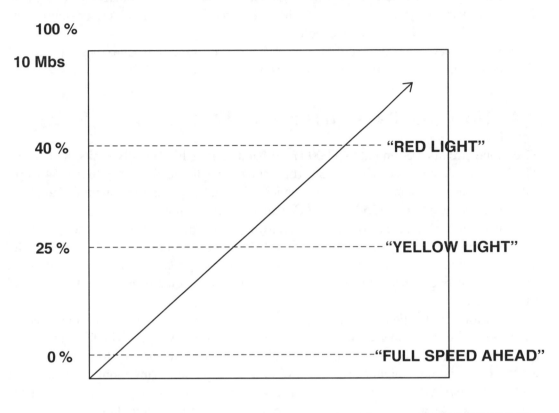

Figure 2.3: LAN throughput.

the price of Fast Ethernet cards or 100 Mbps has dropped significantly, just in the past few months. Therefore, Fast Ethernet is becoming the medium of choice for many, if not most, LAN managers, because of its cost performance and availability.

2.3   Fast LANs

"Fast LANs" have proliferated in user communities, due principally to reductions in the price of the network interface cards (NICs). Where a Fast Ethernet card cost in excess of $300 just a couple of years ago, today the same cards have dropped below the $50 range. Similarly there have been significant price reductions in 100 Mbps hubs and ancillary equipment,

To the typical user, using a Microsoft suite of products (Word, PowerPoint, and so forth) on a properly engineered LAN, 10 Mbps is usually more than adequate. It is the more demanding user, working on computer-aided designs, intensive graphics, or multimedia that will first embrace fast LAN technology.

Speed, however, is not the only variable. Some users require isochronous capabilities. That means the timing relationship of the bits matters. (See Table 2.1.)

2.4   How Fast Can You Run on a Wire?

Telephone circuits are limited to 3,000 Hz of bandwidth, which greatly limits their speed. The best a modem will do is around 35 Kbps, and even a digital circuit tops out at 64 Kbps. LANs have no telco components, like channel units, that limit their bandwidth. Therefore, they are a *lot* faster, up to 155 Mbps on Category 5 wire, or more.

The chief limiting factor on LAN speed is the wire. The higher the frequency you can run on a wire in Hertz (Hz), the faster the speed. Ordinary telephone cable, for example, will carry about 1 Mhz, or up to several megabits, assuming there are no telco components like load coils or channel units. LAN cabling is much higher quality and capable of 10–155 Mbps speeds.

Additionally, the higher the frequency that is run on a cable, the more of a tendency it has to feed over into other cables running the same frequency. At 3 Khz for a telephone channel, this is easily rectified by the telco by twisting the cable once per foot. Twisted pair LAN cabling is much higher in frequency, so it must contain many more twists per foot (unshielded twisted pair) or have a metal jacket around the conductors (shielded twisted pair). Figure 2.4 shows the approximate carrying capacity for different kinds of wire.

2.5   Are LANs Isochronous?

Even though all this capacity is available on the LAN, remember, LANs are *not* isochronous. This will affect your technical decisions on things like video, multimedia, and voice-LAN integration.

LANs are not conducive to delay sensitive traffic. For example, if a user wants to introduce delay intolerant traffic like voice or video into the LAN environment, your options become limited. Ethernet and Token Ring for example, the two most popular LAN protocols, are not isochronous, and therefore are not suitable for voice or video. This could be problematic if your long-term plans call for integration of voice, data, and video. On the other hand, the standards for ISO-Ethernet and ATM allow for isochronous traffic. We will provide a very brief description of each technology in the following sections. Some of these technologies are useful for delay sensitive data, but most are not. They include the following:

EIA/TIA-568 Horizontal UTP Cable Attenuation/Near End Cross Talk (NEXT) Loss (Worst Pair)			
Frequency (Mhz)	Category 3 Attn/NEXT (dB)	Category 4 Attn/NEXT (dB)	Category 5 Attn/NEXT (dB)
0.064	0.9/—	0.8/—	0.8/—
0.150	—/53	—/68	—/74
0.256	1.3/—	1.1/—	1.1/—
0.512	1.8/—	1.5/—	1.5/—
0.772	2.2/43	1.9/58	1.9/64
1.0	2.6/41	2.1/56	2.1/62
4.0	5.6/32	4.3/47	4.3/53
8.0	8.5/27	6.2/42	5.9/48
10.0	9.8/26	7.2/41	6.6/47
16.0	13.1/23	8.9/38	8.2/44
20.0	—/—	10.2/36	9.2/42
25.0	—/—	—/—	10.5/41
31.25	—/—	—/—	11.8/39
62.50	—/—	—/—	17.1/35
100.00	—/—	—/—	22.0/32
Attenuation: Per 100 M (328 feet) 20°C			
NEXT: >100 M (328 feet)			

Table 2.1: How fast can you run on a wire?

- Fast Ethernet—IEEE 802.3u

- 100VG/AnyLAN—IEEE 802.12

- ISO-Ethernet—IEEE 802.9a

- Fiber Distributed Data Interface (FDDI, like Token Ring in ways)

- Gigabit Ethernet—IEEE 802.3z (Hub-Based Fiber)

- Asynchronous Transfer Mode (ATM)

2.6   Fast Ethernet

Fast Ethernet was developed by Fast Ethernet Forum, and is otherwise known as 100 Base-T (IEEE 802.3u, an addition to the 802.3 Ethernet Standard).

It utilizes the same CSMA/CD access protocol as regular 10 Mbps Ethernet, and runs on Category 3–5 unshielded twisted pair (UTP).

Unlike bus-based technologies, Fast Ethernet is deployed in a hub and spoke configuration with a central repeater such as a LAN switch. (See Figure 2.4.) The users served off the repeater must be within a specified distance of 100 meters, or no more than 328 feet. A typical deployment would include repeaters connected by multimode fiber optic cable, with Category 3–5 twisted pair copper cable to each end user device.

Fast Ethernet is nonisochronous. That means if a user hears that you are upgrading the LAN to 100 Mbps and says, "Great, that means we can do video over the LAN!" you will probably have to burst their bubble. Video is delay sensitive and Fast Ethernet, despite its speed, is generally not suitable for it.

Fast Ethernet can be deployed on two kinds of wire, both a four wire and an eight wire configuration. What is really happening here is the specifications allow a user to make up for shortcomings in his or her wiring by using more wire. Cat 5, for example, will easily support 100 Mbps on two pairs. Cat 3, however, will not. So what is a user to do if their building is jammed full of Cat 3? Sometimes in order to salvage their investment, a user will opt for a 100 Base T4 network card. 100 Base T4 essentially sends 33 Mbps on three pairs (six wires) rather than the higher 100 Mbps speed. Since 33 Mbps times three pairs equals 100 Mbps, the user is able to beat the bandwidth problem by simply using more wires. (See Figure 2.5.)

The two Fast Ethernet standards are therefore:

100 Base-TX or, 100 Mbps over 2 pairs
100 Base-T4 or, 100 Mbps over 4 Pairs

The deciding factor is simply what kind of wire you have, how much of it you have, and how willing you are to part with your embedded investment.

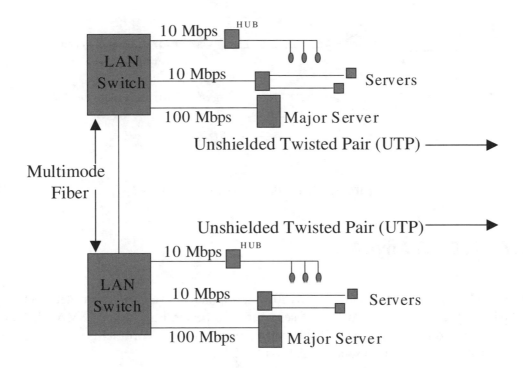

Figure 2.4: Fast Ethernet. A typical 10 Mbps LAN will probably find that as traffic increases, the first segments to become congested are the links between LAN switches or bridges, and other "high traffic" paths such as those connecting to major services.

There are two standards, depending on what kind of wire you have. 100 Base TX: Provides 100 Mbps over 2 pairs (4 wires) of Category 5 wire.

BUT

What if you do not have Cat 5? Then you can use more wires! 100 Base T4: Provides 100 Mbps over 4 pairs (8 wires) of Category 3 wire.

LIKE THIS:

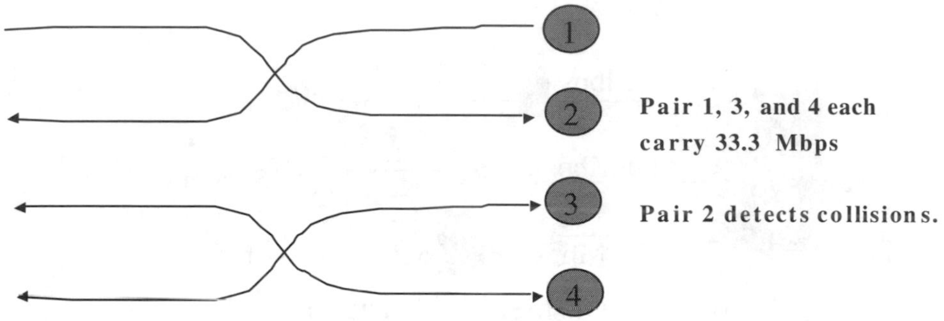

Pair 1, 3, and 4 each carry 33.3 Mbps

Pair 2 detects collisions.

Figure 2.5: How does Fast Ethernet work?

2.7   100 VG/AnyLAN

The 100 VG/AnyLAN standard is IEEE 802.12, originally developed by Hewlett Packard. Like Fast Ethernet, 100 VG/AnyLAN uses a central repeater. Rather than CSMA/CD however, it uses a different process called *demand priority*. Demand priority is something like the "request to send clear to send" signals on RS-232.

Wiring is an 8 wire, 4 bit parallel, half duplex transmission. This allows 100 VG/AnyLAN to act something like 100 Base T4, and use less expensive Cat 3 cable.

Demand priority has two modes. The first is a *control mode* that uses two pairs of wire upstream, and two pairs downstream. The second is a *transmission mode* that uses all four pairs, one way or the other. The stations request permission to send by generating one of two bit patterns (Normal or High Priority) which are also known as "tones."

100 VG/AnyLAN is bilingual. The LAN protocol can be Ethernet or Token Ring once permission is granted, hence the name, 100 VG/*Any*LAN.

Although an interesting LAN technology, 100 VG/AnyLAN seems to be taking a back seat to Fast Ethernet, due to the relatively low cost of Fast Ethernet cards, and further evidenced by Hewlett Packard's acceptance of the Fast Ethernet standard.

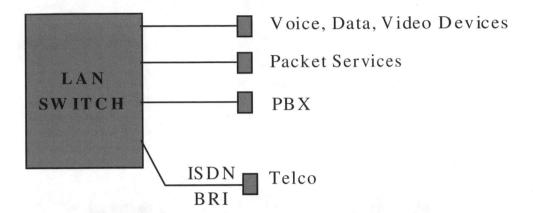

Figure 2.6: ISO-Ethernet. Definitely worth a look! Provides 10 Mbps "P" Channel—*plus*—96 ISDN B Channels Integrated isochronous services for a LAN environment? Perhaps, but can ATM be far behind?

2.8   ISO-Ethernet (IEEE 802.9a)

As the name implies, ISO-Ethernet is isochronous and therefore is suitable for delay intolerant traffic like voice, video, and some kinds of multimedia traffic. See Figure 2.7 to compare isochonous/nonisochronous. In fact, ISO-Ethernet may be the first integrated services interface for the LAN environment. It is cheaper than ATM and available now. Voice, data, and image traffic can ride on the same interface.

ISO-Ethernet uses 2 pair EIA Cat 3 or 5, and similar to other LAN technologies has a 100 meter range from the hub. The hub is usually a LAN switch referred to in the vernacular as an access unit, or AU.

As shown in Figure 2.6, ISO-Ethernet operating modes include:

- **A 10 Base-T Mode:** Provides 10 Mbps P Channel only

- **A Multiservice Mode:**

    - One 10 Mbps Packet Channel (P)

    - 96 64 Kbps ISDN Channels (C)

    - One 64 Kbps ISDN D Channel (D)

- **An All Isochronous Mode:** The standard also defines a configuration where the entire capacity is used for C Channels and delivers 248 64-Kbps channels.

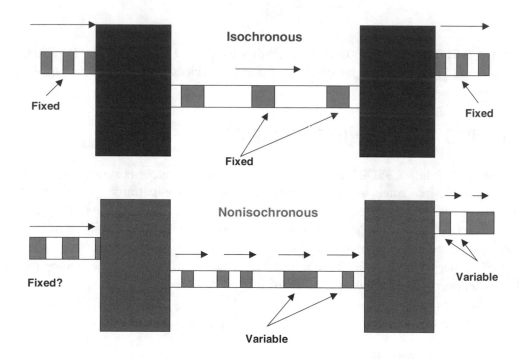

Figure 2.7: Isochronous/Nonisochronous.

2.9   FDDI (Fiber Distributed Data Interface)

Is FDDI a "legacy" LAN technology (see Figure 2.8)? Apparently so! The ANSI X3T9.5 (1989) standard that defined it is many years old. That also means that it is widely supported and thoroughly run in.

FDDI is used primarily for backbones and campuses. It is internally much like a "souped up" Token Ring, and operates much the same way. It is designed to operate on single or multimode fiber up to 100 Km (60 miles).

Like Token Ring, FDDI allows one device to send at a time, and only if the device has the "permission slip" or token. There are no collisions like in the Ethernet CSMA/CD world. FDDI does however hold the token longer (5–20 Ms) than Token Ring.

FDDI maps easily to SONET, which makes it relatively straightforward for WAN extensions. It does use a 4B for 5B (Four Bit for Five Bit) type of encoding which amounts to a fairly significant overhead. That means you really need 125 Mbps of SONET capacity to map 100 Mbps of FDDI, since you only get four bits of payload for each five bits sent. In other words, a 155 Mbps OC3 is the ticket.

Although FDDI is a mature, reliable technology, it is gradually giving way to switched LANS and other more modern technologies.

2.10   Gigabit Ethernet

If Tim Taylor of television's *Home Improvement* was a LAN Manager, what would he use? Why, Gigabit Ethernet of course—LAN technology for real men. (This works okay, just needs more power, right?)

Seriously speaking, Gigabit Ethernet provides a *billion* bits per second (well, okay, really "only" 800 Mbps due to signal coding) but only for short distances with today's technology. Although the standards are still coming together, we do know this much. Gigabit Ethernet is based on the IEEE 802.3z standard. It is designed to be deployed to a large server or high speed backbone.

The specifications allow for multimode fiber connections up to 500 meters, but as I mentioned, 8B/10B line coding means the line rate must be 25 percent faster, just like in the FDDI example above.

One of the problems with Gigabit Ethernet lies with its range. Gigabit Ethernet still uses the same CSMA/CD technology. The problem is, collisions take on a whole new dimension at a billion bits per second. Do not worry—nothing explodes. But since the signal propagates so quickly, a collision can actually occur on the line after the network interface has finished sending—meaning it never knew that it happened.

At 10 Mbps, this does not present much of a problem since the furthest two devices on the network can be well over a mile apart. Even at 100 Mbps, the distance is still a halfway respectable 200 meters. But at 1 Gbps, the distance drops to 20 meters. So what

(ANSI x3T9.5—1989)

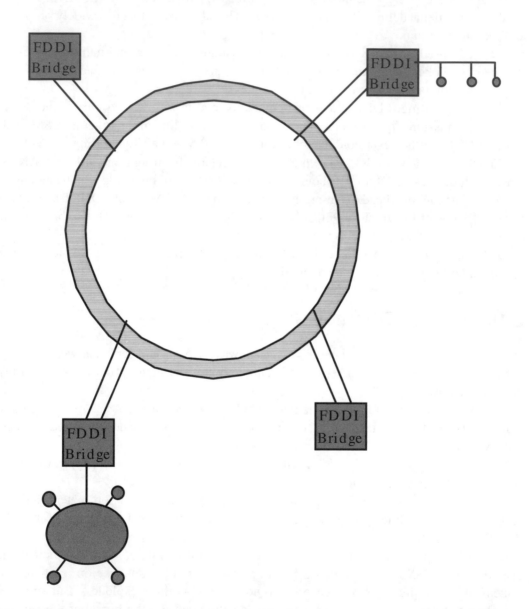

Figure 2.8: FDDI—A "legacy" LAN technology? Sure looks that way. Standards have been around since 1989. Widely supported, but gradually losing ground to other technologies.

At 10 Mbps	64 Octet Frame Takes	51.2 Microseconds (USEC)	2000 Meters
At 100 Mbps	64 Octet Frame Takes	5.12 Microseconds (USEC)	200 Meters
At 1 Gbps	64 Octet Frame Takes	.512 Microseconds (USEC)	20 Meters

Table 2.2: Gigabit Ethernet range problem.

kinds of connections can be made that are less than 60 feet long? Well, some backbone connections between high speed LAN switches may be in that range. So can high speed channel connections between different kinds of "legacy" equipment. Most likely, the first Gigabit Ethernet link will be only 19 *inches* long—the distance from one LAN switch to another in a standard 19-inch cabinet. (See Table 2.2.)

2.11 ATM

Why do LANS and WANs have different rules? ATM is both a LAN and a WAN technology. ATM is a high speed cell switching technology. ATM is isochronous and interfaces to delay intolerant traffic. ATM will support voice, data, or video. ATM also operates at a number of speeds, including:

- T1/1.544 Mbps

- IBM Proprietary/25.92 Mbps (which is exactly half the OC1 line rate of 51.84 Mbps)

- T3/44.736 Mbps

- OC1/51.84 Mbps

- FDDI/100 Mbps

- OC3/155 Mbps

- OC12/622 Mbps

- and others

ATM will eventually replace the other LAN technologies when users need a combination of isochronous capability and high capacity. It is also the most flexible way to apportion SONET traffic, as we will illustrate in later chapters.

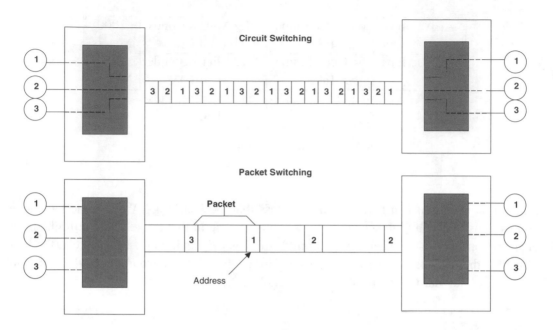

Figure 2.9: Circuit switching versus packet switching.

Figure 2.9 shows the difference between circuit switching (where everyone gets an even split of the circuit) and packet switching (which is essentially all you can grab on a first-come first-serve basis). There are no precast time slots in packet switching. This makes it efficient. If user 1, for example, stays home from work. users 2 and 3 can share his capacity. In circuit switching, user 1 ties up his circuit whether he is in the office or not.

Figure 2.10 introduces *cell switching*, which is really a souped-up version of packet switching. Cell switching is the basis for ATM (Asynchronous Transfer Mode). Cell switching has the same idea behind it as packet switching. It allows users to efficiently share capacity. Cell switching (ATM), however, is packet switching on steroids. It provides for greater speeds.

For the time being though, who will be the most likely early adopters of ATM? The answer is users with either extraordinary bandwidth requirements, like those manipulating 4 to 8 Giga*byte* chip diagrams for example, or users who need a combination of both high capacity and isochronous bandwidth. Figure 2.11 shows an actual example of an engineering application. The company pictured was an early adopter of ATM technology, beginning their deployment in 1993!

The company pictured today has an OC48 (that is about 2.4 gigabits per second!) backbone—and growing pains!

Consider the effect however. A 4 Gb chip diagram which used to take 168 hours to transmit using the highest speed modem available, and almost an hour and a half even with

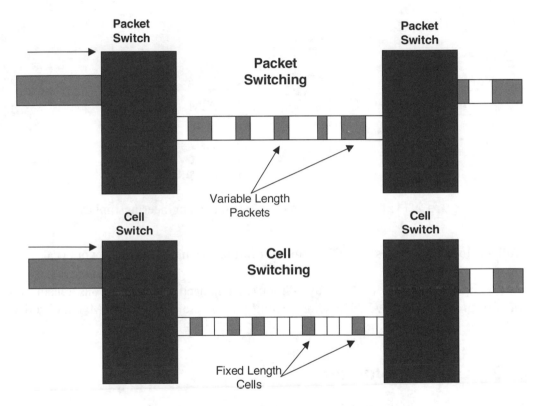

Figure 2.10: Packet versus cell switching.

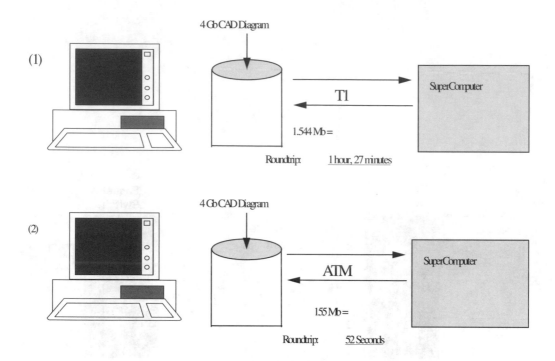

Figure 2.11: Effect of ATM technology on an engineering application.

a full T1, today now takes only 52 seconds. Now that is what Tim Taylor would call "more power!"

If you serve a stable of "power users" or users that need true isochronous transmission, you will probably be an ATM candidate shortly if you are not already. More ATM detail will follow in Chapter 6.

2.12   LAN Topologies

The following figure illustrates several LAN topologies. Watch later in the book as we introduce many of these concepts (ring, mesh, and so forth) in the context of the Wide Area Network as well! (See Figure 2.12.)

2.13   Multimedia Applications

Imagine that after several days of house hunting you and your spouse find the perfect dream home—just the right neighborhood, just the right schools, perfect size, and hopefully perfect price. The real estate agent offers to bring you back to her office where she can pre-qualify you to see if you can afford the mortgage and hopefully avoid an embarrassment

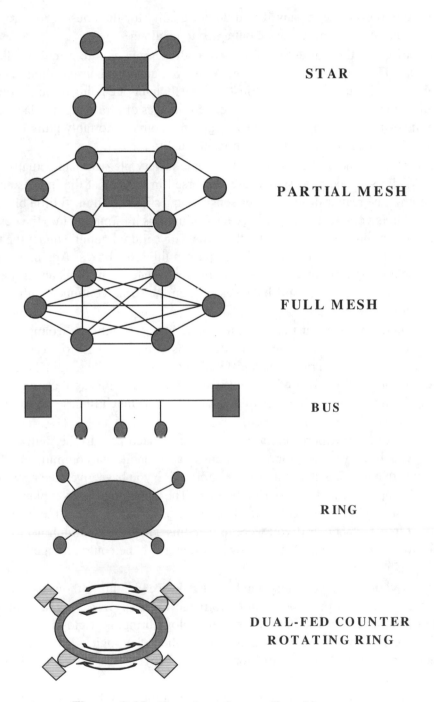

STAR

PARTIAL MESH

FULL MESH

BUS

RING

DUAL-FED COUNTER ROTATING RING

Figure 2.12: Overview of network architectures.

later with the mortgage company if you do not qualify for the house. You are somewhat skeptical and a little bit apprehensive but agree to go along.

Upon arrival at the real estate office, the realtor brings you to a multimedia terminal device situated in the corner of her office. Again you are apprehensive about handing over personal financial data to a realtor, but just as these thoughts are going through your head the realtor fires up the device and suddenly there are images of birds singing, dad cooking hot dogs on the grill, kids swinging on the swings, and mom comfortably lying in a hammock in the beautiful dream home which you hope to qualify for.

Music plays in the background and the scene is very pleasant and actually quite captivating. While you and your spouse are immersed in looking at the images and smiling at one another the realtor drops a rather sensitive question. "Tell me a little bit about your income. What is your annual salary?" After feeding this information into the device, more pleasant images and music come forth, this time with children running under the Christmas tree, warm fireplaces, and puppies barking up from little shoeboxes. Again, while you and your spouse are captivated with this scene, the real estate agent turns and says, "Tell me a little bit about your credit card balances." Then the scene continues with more happy images of home ownership.

Once again I'm being just a little bit tongue in cheek with the example. But when asking people about delicate financial matters, it is oftentimes best to create a little bit of an ice breaker or diversion rather than just sitting down at a sterile terminal asking probing personal questions. The intent of the multimedia terminal in this application was to make qualifying for a mortgage a fun experience, or at least as much fun as one can have under the circumstances.

At the end of this experience, a color printer situated next to the multimedia device prints out what is known in the mortgage industry as a conditional commitment. The conditional commitment means that if all of the information you have given the agent is correct, Mortgage Company X will underwrite the loan. The agent goes on to explain that you are free to choose any mortgage company you want, of course, but that the experience will probably go much smoother if you were to take this conditional commitment to Mr. John Smith at Mortgage Company X tomorrow. Or better yet she could arrange for him to call you.

Now ask yourself, do you think that having a device like this in thousands of real estate offices might generate some money for Mortgage Company X? Do you think this is an example of taking a product and through technology bringing it closer to the customer? Do you think that this capability would be worth the technological investment? That this investment would pay off in sales for Mortgage Company X? The answer to all four of these questions is a definitive yes.

However, what you will find is even more of a surprise. What I have just described is a 1985 application. A mortgage company I worked for some years ago devised the system called RMO, or remote mortgage origination, and the idea was exactly what I have just described: to put multimedia devices in real estate offices in order to bring a product which is normally a fairly hard sell, for example a mortgage, to the place where the action occurs and where people buy houses. The idea was so good, in fact, that there was even discussion internally that this might constitute restraint of trade. Very similar to the situation in which an eye doctor locates directly across the street from an eyeglass clinic. You will notice in the eyeglass clinic that they always tell you, "You are free to use the optometrist of your choice, however there is one across the street for your convenience."

It is a very similar process here. What is almost comical to look back at now was the type of equipment we had to use for this solution in 1985. The multi-eight media device was a 386 PC with an EGA monitor. It came with a video disk player the size of a coffee table and carried a hefty list price in the neighborhood of $35,000. In other words, we used the equivalent of stone knives and bear claws in comparison to what could be bought today for a similar application. In fact, the device probably could be deployed today in the $2,500–$3,000 range using off-the-shelf PC technology and MPEG video graphics, CD ROMS, and other components.

The important thing that I am illustrating here is the idea or, as I stress repeatedly in this book, the way that a strategic use of technology to bring your business closer to the customer grows and creates killer applications and does some marvelous things for your business. To plan capacity for multimedia, refer to Table 2.3.

2.14   Engineering "Killer Apps" Through Seamless Solutions

You will recall our explanations of how seamless solutions strengthen the organization as a whole and equate to higher levels of sales or customer confidence. It is very easy to draw a series of concentric circles, like we have in Figure 2.13, and pretend that these represent something close to real life. Real life, however, is more in line with Figure 2.14. What you see is that even though the circles overlap on the systems level, there is a lot of physical distance between locations which needs to be overcome by network technology in order to bring these seamless solutions to life. For example, the customer support center or customer service in Figure 2.14 is in Newark, New Jersey. The sales organization is in New York. Engineering is in Dallas. Production as well as shipping and receiving are at a manufacturing plant in Monterey, Mexico where labor costs are cheaper. Post sales are in New York, as is the accounting function.

When one considers Figure 2.14, many different types of technologies come to mind. Would the planner consider a front-end processor in New York where presumably the mainframe still resides? Would routers be used in Dallas, or would they be shunned in favor of LAN switches? LANs transfer data faster which is more to the liking of engineers, who have to move large computer aided designs. Would a T3 or an OC3 make more sense

SERVICE	MEDIA	FILE SIZE	DELAY TOLERANT?
Text, 3270 Terminal Emulation	LAN, COAX, WAN	1K Per Transaction	Yes
FAX Transmission	Dial-Up Voice Grade	10–25KB Per Page	Yes
Internet—Web Page Access	Dial-Up Data, ISDN, T1	100–200K Per Page	Yes
Image (JPEG Compressed)	LAN, WAN, Dial-Up, Frame Relay, Other	50–500KB	Yes
Image (Photo Uncompressed)	LAN, WAN, Dial-Up Frame Relay, Other	1–7MB	Yes
X-Ray Image	LAN, WAN, Dial-Up Frame Relay, Other	7MB+	Yes
Human Voice	PSTN, WAN	64Kbps	No
Video (Compressed)	ISDN BRI, Switched 384, T1	128Kbps–1.544Mbps	No
Video (Studio Quality)	T3-OC-XX SONET	44.736–155Mbps	No
HDTV	OC-XX	155Mbps+	No

Table 2.3: Planning capacity for multimedia.

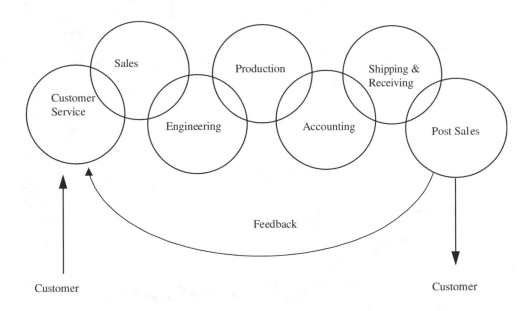

Figure 2.13: Using LANs for a seamless solution.

between engineering and production, giving the company the capability of moving large computer designs back and forth between engineering and the production mechanism at the factory? Do T1 links between the shipping and receiving department and the accounting department make sense for seamless accounting and tracking of shipments, billing, and other data? Once the business plan is hammered out, then the technology overlaying the process becomes easier with cost justification being simple arithmetic. For example, adding up the prices of all the boxes and telecommunication links required to bring this solution to life is simple although the sum will be a large number. Therefore it is good to have something on the benefit side of the equation allowing you to make profound statements such as, "We propose to spend 1.4 million dollars on new LAN and WAN technology and as a result we expect to derive 14 million dollars in cost benefit in terms of increased sales and reduced overhead expenses such as personnel duplication." You would be surprised how many companies start with only a network diagram like this one, and actually expect to close the sale.

Human beings being what they are expect to have some type of reward in return for spending money or dedicating resources to a project. Therefore, it is important to present the benefit slides to executive management first. And only after the benefit of the solution is firmly implanted in management's mind should the technologist lay out technical diagrams and associated costs.

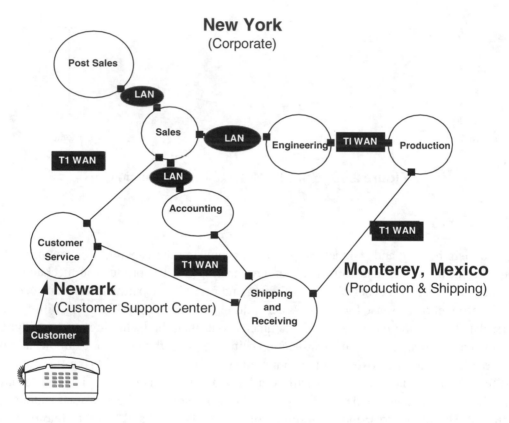

Figure 2.14: Seamless solution—the 100,000 foot technical view of a hypothetical manufacturing company.

2.15 Summary

Once again this chapter was not designed to be a definitive dissertation on LAN technology, moreover it was intended to allow graceful integration of LAN and WAN technology by providing a framework of reference in the mind of the telecom planner. As years progress, the distinction between LANs and WANs will continue to diminish, especially in light of upcoming technologies such as ATM, where the distinction between LANs and WANs vanishes entirely. Clearly the benefit to LAN technology is in being able to foster innovative seamless solutions such as these.

By being able to integrate the appropriate WAN and LAN technology, your organization will be able to extend these efficiencies to outlying locations, positioning itself for greater efficiencies, higher sales, and better integration.

The next chapter will talk in greater detail about the wide area network services which link your disparate locations together into an integrated seamless solution.

3

WAN Services Overview

3.1   Tying It Together

The term telecommunications comes from the ancient Greek or Latin (I do not remember which) but it means roughly to communicate from afar. Therefore, it seems rather preposterous that we would tell our users that we would like to give them business functionality, but that they have to do it here.

The role of the modern telecommunications planner is to extend capabilities over a distance, creating new efficiencies and avoiding duplications in the organization. The seamless solutions we described in Chapters 1 and 2 are very hot concepts and become even hotter when files and resources can be shared not only in the local environment but also between widely disparate company locations.

This chapter will introduce the various products and services that your organization can utilize to bring these capabilities to life.

3.2   The WAN Environment

In Chapter 2 we discussed the LAN environment which, colloquially speaking, means everything "in here." Therefore, it makes sense to assume the WAN environment is everything "out there." The problem is as soon as you extend your business "out there," you lose control of what goes on. In fact, items such as telecommunications contingency planning and disaster recovery have taken on new importance over the last few years.

For example, consider the *Home Shopping Network*. An item is displayed on the television screen and callers call in via an 800 number to an agent tied to a local area network "and a seamless solution" to order the beautiful Armenian Gold flatware set currently on the screen. What do you suppose happens to *Home Shopping Network* when Billy Bob and his big back hoe dig up the fiber optic cable serving the telemarketing center? For all practical purposes, the cash register stops. Even more distressingly, businesses like this depend on fast turnover of product—clearing things out of the warehouse today to make room for new things, which will be in the warehouse tomorrow. Provided of course that they ever make it to the warehouse at all.

It becomes very apparent how significant a telecommunications disruption can be in this type of environment. And for this reason, we will also make mention of various types of contingency planning and backup systems, which should be in place when proposing any type of network-centric solution. For the time being however, let's talk about the WAN.

Flip ahead a few pages in the book to Figure 3.1 which illustrates various components of the wide area network. You might want to mark the page with your finger and refer to it while reading the text on the following pages. The dark portions of the figure indicate components that are traditionally considered wide area network or WAN components. These include, of course, T1s as well as their associated equipment called DSUs and CSUs. Just a few short years ago, the DSU and CSU were contained in two separate boxes. The DSU took care of taking the digital signal that your equipment speaks such as a personal computer front-end processor or channel bank and converting it into the digital signal that the phone company speaks. In some ways it is almost like a modem, except a modem takes a digital signal from your computer and converts it to an analog waveform that can be sent over a phone line. A DSU takes a digital signal from your computer and simply converts it to a different type of digital signal that the telephone company recognizes called a bipolar digital signal. We will discuss this in the T1 section in Chapter 5. The CSU provides a protective function. For years, the telephone company had a lot of trepidation about people attaching non–telephone company devices to the network. They believed that protective features were necessary to protect the public network from devices over which they had no quality control. The CSU also selects something called a line code in the T1 world. There are two prevalent line codes called AMI (alternate marked inversion) and B8ZS (bipolar 8 with zero substitution). These are also discussed later in Section 5.12. As the result of both regulatory decisions and advancements in technology, the DSU box and the CSU box were eventually combined into one box called (surprise!) DSU/CSU. And today the state of the art is almost to the point where a DSU/CSU can virtually be contained on one chip.

Another part of the wide area network is the channel bank, which we discussed earlier. Although a channel bank can reside at a customer's location, it was originally designed as a piece of telephone company equipment. We will discuss several other pieces of equipment during the course of this book which originally started out in the telephone company's network, but which are now common fixtures on the premises in large companies. SONET PTE (Path Terminating Equipment) for example, is one of these.

Looking over to ⑦ in Figure 3.1 you will see an ATM switch. We have divided the ATM switch in half in the figure to illustrate the point that ATM is really both a local area and wide area network technology. ATM in turn maps very nicely over SONET (⑧ in the figure). The SONET PTE shown in the figure is usually at the telephone company, but again as technology and regulatory constraints continue to change, it can also be located on a customer's premises. In the middle of the figure by ⑤ is the Frame Relay network, shown by that infamous cloud. Everything from the Frame Relay access device (FRAD) outward through the cloud is also considered the wide area network. The FRAD again is illustrated by a box divided in half since half of the box speaks the Frame Relay protocol or a wide area network protocol, but the other side of the box probably speaks a LAN protocol or something designated for the local network. Unlike ATM, however, a protocol translation must take place here in order to convert the Frame Relay format to the Token Ring, Ethernet, or IP format, that it must speak out the other end.

Looking over to ⑩ in the figure, the Internet itself can be considered part of the wide area network. The Internet in the figure is illustrated again by a cloud. Inside the cloud, however, are thousands of independently intelligent and independently connected servers, routers, bridges, LAN switches, and telecommunication links. Even though, technically speaking, the Internet can be called a WAN in today's corporate vernacular, it is usually referred to as the Internet rather than part of the WAN and is recognized as a separate entity apart from the corporate communications WAN network. Looking down just below at ⑮ in the figure are ISDN BRI and PRI. These are also wide area network technologies.

3.3   What Are Modems?

The word *modem* is a contraction of two words meaning modulator demodulator. As we stated earlier in this chapter, a modem takes a digital signal from your PC and changes it to a format that can be transmitted over an analog telephone channel. Ask yourself what were telephones designed to do? Obviously—to talk. A human voice over a telephone wire is really a continuously varying wave form of electrical impulses carried in an analog format, which at the far end is translated from electrical impulses to sound by a vibrating speaker. A digital signal, however, is different. Consider this example. I walk over to a light switch in your office and I turn on the light. (Say this is a one.) I turn off the light. (Say this is a zero.) Now with a series of ones and zeros, I am going to send you a message in a secret code. Every second, I am either going to turn the light on or I am going to turn the light off. You, in turn, are going to watch the second hand on your watch to determine if that particular second was a one or a zero. Furthermore I will give you the code. Light, light, light, nothing, nothing, nothing, light, light—that is the letter A. The next segment of the code is light, light, light, light, nothing, nothing, light—that is the letter B. (They aren't really, but go along on me with this one.)

Using this type of code, I can actually flash intelligent messages, and since you are a very intelligent and astute person, you can keep up with me. Imagine, however, that you have to run out for a moment for a cup of coffee. When you come back, I am in the middle of blinking the light. Now, you do not know where I am in the code and you have immediately lost your place. In other words, you have had a frame slip just like your T1s do. Therefore, every so often I have to say the words, "ready, set, go," and at that time you know I am beginning the sequence and you are back in sync with me again. Now we are synchronized just like your T1. Okay, if we can do this with light we ought to be able to do it with batteries. Let's assume you and I are going to work another secret code out between ourselves, only instead of using the light, battery is a one, no battery is a zero. Once again you will watch the second hand on your watch. I will give the command, "ready, set, go," and then I will send you a series of electrical impulses: battery, battery, battery, battery, no battery, no battery, battery, battery—there is the letter A again. Once again since you are an intelligent and astute individual, you have no problem with this code as I flash you message after message. If you are with me so far, we can actually take this another step. What if I send you a positive voltage and call that a one, then I send you a negative voltage and I call that a one, and when I send you no voltage that is a zero. Let's try that again: positive battery, negative battery, positive battery, negative battery, no battery, no battery, positive battery, negative battery—there is the letter A again. Only this time, I am using a bipolar digital signal meaning alternating pulses of positive and negative just like the telephone company uses on your T1.

Okay, now that we have introduced the concept of a digital signal and you have a little bit of an idea of what I am talking about, let's get back to modems. Since telephone channels are designed to carry voice frequencies and designed to carry varying analog waveforms, it is impossible to send light or batteries over a standard telephone line. Therefore we must take these digital signals and convert them to sounds, which can be sent over the line and reconverted to digital signals on the other end. This is especially silly considering the fact that 95 percent of the public telephone network is digital.

However, the copper cable between the central office and our homes or businesses is not. It is analog and it is because of this 18,000 foot length of cable that we go through all the trouble of having modems. So once again, modem stands for modulator demodulator and it takes the digital signal from your computer or CPE which comes in the format of battery, battery, nothing, nothing, battery, battery, and it converts it into tones which can be sent over a phone line. Can you send digital signals over a voice circuit? I suppose you could on some, but it would be akin to listening to a flashlight. That is why we use modems. In essence, modems take digital information and turn it into analog. Just to confuse things more, another device called a Codec or coder decoder, takes analog and converts it to digital. For example, a Codec will take an analog signal from a television broadcast and convert it to a digital signal, which can be sent over a T3. But rather than confuse you with that now, let's move on.

Table 3.1 illustrates different types of dial-up modems and the year in which they were introduced. Virtually all modems shown in this figure are technically or functionally obsolete. In fact, it is arguable that all modems will become obsolete as digital technologies become more prevalent and as other superior technologies such as ISDN make their way into homes and small businesses. We are also hitting the end of the line with regard to modem speed. Most experts agree that the theoretical limit for a symmetrical 3 kHz dial-up modem lies at about 35 kilobits per second. With today's technology being in the neighborhood of 33.6 Kbps on a symmetrical connection it could be argued that we have essentially wrung every bit of bandwidth we are going to get out of a standard telephone channel.

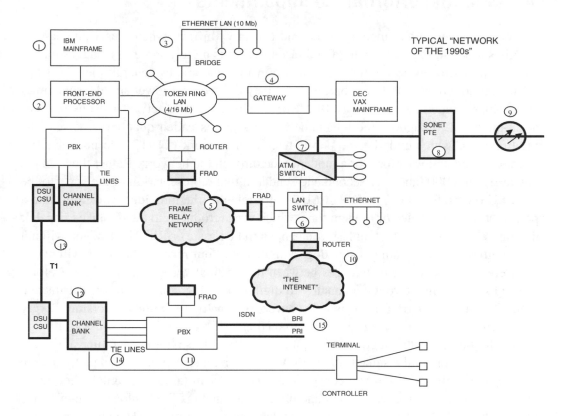

Figure 3.1: The WAN environment.

Type of Modem	Speed	State of the Art in:
BELL 103	300 bps	1968
BELL 212	1200 bps	1977
ITUV.22	2400 bps	1983
ITUV.32	9600 bps	1989
ITUV.32bis	14.4 Kbps	1990
ITU V.34	28.8 Kbps	1995
ITU V.34+	33.6 Kbps	1997
"End of the Line"	35 Kbps*	
* As far as we can tell today, 35 Kbps is the theoretical limit for symmetrical 3 kHz dial-up modems.		

Table 3.1: Kinds of dial-up modems.

3.4   Modem Performance Impairments

A variety of factors affect modem speed and capacity. In fact, when a user purchases a 33.6 Kbps modem, that is not a hard and fast speed, it is merely more of a suggestion. A variety of impairments on the line can affect transmission, and as general rule 33.6 Kbps modems do not connect at 33.6 Kbps. Instead they connect at a lower speed—sometimes 24,000 Kbps, sometimes as low as 19.2 Kbps. There are a variety of reasons for this. One is prevalent throughout the telephone network and is called quantizing noise. The digital carriers, which underlie the public telephone network (T1–T3), do not send 100 percent of an analog waveform or a modem tone through the network. Rather they sample this waveform 8,000 times per second and send a digital representation of the wave to the DSU/CSU on the far end. The DSU/CSU in turn rebuilds the individual waveform into a representation of the voice or modem tone, which was introduced to the digital carrier. For voice this works great. In fact, unless you happen to be Lassie or Rin-Tin-Tin, you will not hear the difference, and I am not sure dogs could either. Your modem, however, will notice the difference. The subtle differences between the original waveform and the waveform that results after going through an analog to digital to analog conversion, are detrimental to data. The industry term for this is A/D conversion (analog to digital conversion). Every time a signal goes from analog to digital, to analog to digital, it degrades just a little bit, in much the same way that a good office joke degrades as it is photocopied time after time after time. It is kind of the same thing with A/D conversions. A typical modem to modem telephone call will have two A/D conversions. The modem takes the digital signal from your computer or customer equipment and turns it into analog. The analog signal in turn is sent over an 18,000-foot telephone line to a telephone central office. The phone company then has a codec in the central office that takes the analog signal (the codec is inside the channel unit that resides inside the channel bank) and introduces it to a T1. The T1 in turn, being designed for interoffice transport, transports the voice or data call to a central office closest to the far end of the data or voice circuit. At that point a codec in the channel unit in the far end converts the digital signal into analog. Word is sent to the modem on the far end. The modem, in turn, takes the analog signal and converts it into digital to talk to the other computer. Therefore the number of analog to digital conversions which occur here is two. Fifty-six Kbps modems get around a little bit of this problem by allowing only one analog to digital conversion. This will be discussed in Section 3.5. Other performance impairments on analog circuits include:

- Frequency Response—telephone lines pass some frequencies more easily than others. In general, the frequency range between 1,000–1,800 Hz is attenuated less through a telephone line than frequencies at the high and the low end of the spectrum (for example, the 300–500 Hz frequency spectrum and the 2,500–3,000 Hz frequency spectrum). When the response of the line is not completely flat this can wreak havoc with data transmission.

- Channel Noise—one of the amazing things that has occurred in the last 10 years in telecommunications is the improvement in quality of communication channels. Even 10 years ago a common noise reading for an intercity telephone channel was 45Dbrnc0. Dbrnc0 means decibels of reference noise with reference to zero. Today, a similar reading would be somewhere on the order of 19Dbrnc0. With the advent of clean, quiet digital transmission facilities based on fiber optics, one need not even measure for noise levels in most cases today. In fact, it is commonly held that a typical dial-up long distance telephone call today is far superior in quality to the finest condition private line data circuit even just 10 years ago.

- Delay Distortion—some frequencies travel faster through a telephone line than others. The fastest frequency through a telephone line is approximately 1,800 Hz, with the slowest being at the low and the high ends of the spectrum. This, however, is less of a concern than it used to be, since most modem delays equalize themselves. When you hear your modem connect to the Internet every evening, and it makes that sound like a vacuum cleaner gone crazy, it is really doing an instantaneous delay and frequency check of the line to decide what the characteristics of the line are and how fast it can run.

- Other Impairments—since the network infrastructure serving homes and small businesses is still based predominately on copper cable, a number of impairments can be problematic. Ground hums, noise, electrical interference, or (at the extreme) lightning strikes can all wreak havoc on data transmission using modems.

So if the world is going digital, like this book has presupposed, why then do we devote space to antiquated subjects like delay distortion and circuit conditioning for analog services? Well, like the old saying goes, don't ever forget an old trade because in time everything comes back around. Ever hear of ADSL (Asymmetrical Digital Subscriber Line)? Well, the name is a misnomer. ADSL is really an analog modem signal, but it is really like a telephone channel on steroids. It passes a much higher frequency and thus runs far faster than a telco channel. Being analog however, it is susceptible to the same kinds of impairments as those just described. Just when we get rid of delay equalizers and data scopes, it might be time to drag them out of storage again! ADSL is described in greater detail later in this book.

3.5   56 Kbps Modem—Poor Man's ISDN

Another way advancements in general network technologies have also advanced modem technology is through the recently introduced 56 Kbps modems. First let's just dispel a couple of misconceptions about 56 Kbps modems.

The modems are not really 56 Kbps, but are actually 53 Kbps due to FCC regulations, although we have probably seen this on the box many times. The reason for this is that the initial 56 Kbps modems introduced actually tried to steal a little bit of extra bandwidth above 3,000 Hz and below 300 Hz. Since these are reserved for guard bands to prevent interference with other channels, the FCC looked and instructed the modem manufacturers to limit the bandwidth of the modems to the frequency spectrum between 300 Hz and 3,000 Hz. This limited the speed of the modem to 53 Kbps, which is the present state of the art.

56 Kbps modems are not 56 Kbps in both directions. Generally speaking, a 56 Kbps modem gives the user 56 Kbps downloads, but only 28.8 Kbps uploads. This makes the modem ideal for Web surfing when it is generally more important to the user to download a big file rather than upload a big file. Since Web surfing, for example, involves only clicking and small data transfers upstream, but big, slow moving Web pages downstream, these modems make sense for these types of applications. In telecommuting applications, however, the intent is generally to download a file, manipulate it, and then upload the file back. This means that even with a 56 Kbps modem, the user will be pleased with the downloads, but generally underwhelmed with the uploads.

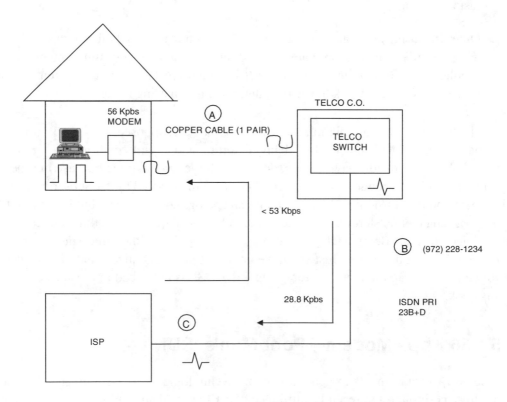

Figure 3.2: 56 Kbps modem—"poor man's ISDN."

56 Kbps modems are not 56 Kbps. The speed again is a suggestion and not a hard and fast rule. In fact it is more the exception than the rule that these modems run at 56 Kbps. Generally, they run at some speed less due to impairments on the line. If a user lives too far from a central office, or uses a central office that does an extra A/D conversion, or is served from a central office or remote unit using anything less than state-of-the-art technology, the result will almost unquestionably be a speed less than the 53 Kbps for which the modem is rated. Figure 3.2 illustrates a 56 Kbps modem connection and the relationship of A/D conversions to this connection. In this figure, the user dials up an Internet service provider and the unipolar digital signal from his PC is converted to an analog waveform by the modem. The signal remains analog (A) from the user's house up to 18,000 feet to a telco central office and a telco switch; the switch has a card which converts the analog signal to digital and introduces it to one channel of a 23B+D ISDN PRI shown as (B). The PRI is a digital facility with 23 telephone numbers assigned to it. The signal stays in digital format all the way to the Internet service provider (ISP) shown at (C). Therefore, instead of being two analog to digital conversions, as would normally happen on a modem transmission, there is only one and the line is cleaned up to an extent where it will support a higher data speed or one up to 53 Kbps downstream and 28.8 Kbps upstream.

ISDN—rather than poor man's ISDN how about real ISDN? As digital technologies become more prevalent in the public network, the last logical place to deploy them is in the "last mile" to small offices, home offices, and small businesses. If you look at Figure 3.3 you will see that on a typical transcontinental phone call, the whole circuit is almost all digital anyway. Upgrading the "last mile" as shown in Figure 3.2 allows the end user to avail himself of the line's full capacity. This is where ISDN BRI is deployed. ISDN BRI provides two 64 Kbps B channels with one 16 Kbps D channel in a format not unlike a miniature T1, and it does so up to 19,000 feet on standard telephone cable. Over 90 percent of the access lines in the United States are at least capable of carrying ISDN traffic with minor modifications.

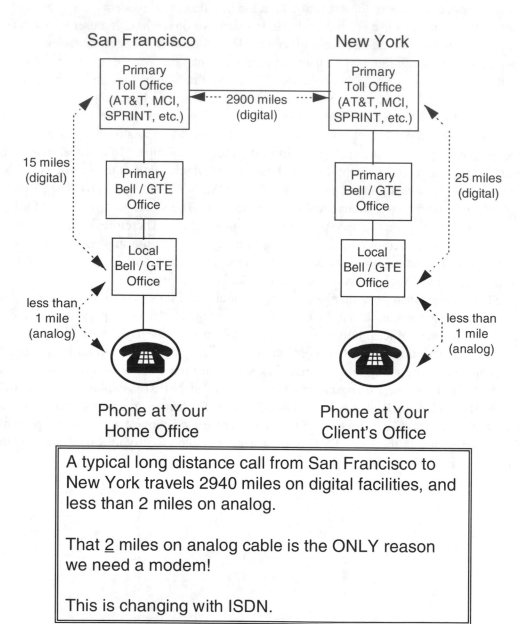

Figure 3.3: All digital network major points.

4

LAN/WAN Interconnection Basics

- How to Make Things Transparent

- How to Sell the Package to your Boss

4.1 Growing Killer Apps

What is the difference between a ho-hum service and someone exclaiming, "Man! I have *got* to get me one of those!" Sometimes things that sell completely defy logic, case in point, Beanie Babies and Cabbage Patch Dolls. Beyond these anomalies however are lots of things which win widespread acceptance in the user community and bring fame and fortune to their purveyors. Another word for these dramatically new business processes? They are commonly referred to in the vernacular as "Killer Applications." And you know what? Forward-thinking companies can actually *grow* them!

Killer applications really do exist, as evidenced by the abundant success stories of companies who applied technology in a manner no one had ever dreamed of before, and rocketed their firms into technological superstardom. Consider the American Airlines Sabre Network, or the way Federal Express can tell you *exactly* where your package is at any given time. These are all adaptations of technology that have effectively contributed to the core businesses' bottom line in "killer" fashion to be sure!

How successful has your organization been in identifying these technological "shooting stars?" What kind of payback to the company has been customary when dollars are invested in technical systems? Can you even measure the effect these expenditures on technical services have on the company as a whole? If not, your chances at technological superstardom are abysmal, and worse, your chances for long term survival may be as well.

Technologists who cannot quantify their contributions to their companies in terms executive management understands (meaning *dollars*) are prime candidates to be downsized, rightsized, capsized, or worse. So how can you, as a technologist, justify your existence to management? How can you identify those "killer" applications which will make you a technological superstar both within your company and externally, among your peers? How can you point to millions of dollars in new company revenue, produced as a direct result of your technical design, and proudly say *"I did that!"*?

4.2   Developing a Strategic Plan

You may already have lots of good ideas about how and where you can improve the technological base that supports your company. Before you begin to draw up advanced plans for lots and *lots* of super-duper systems, you should make sure that your business case is written in such a way that *your* bosses will sign off and endorse it. In order to buy off, and fund your ideas, they will first have to understand them.

Rather than dancing up and down talking about the virtues of technology, specific business solutions must be packaged into winning combinations of services. These must concentrate on the core business and make specific business problems go away. They must also be communicated in a manner that is understandable to the senior management.

Remember, *business process improvements* sell, not technology improvements. The following make great hot buttons when identifying these business processes. Applications which do one or more of the following are the most sellable to executive management:

- Shorten product cycle time

- Bring the customer closer

- Provide better customer service

- Provide ease of use of your product

- Make employees more productive

- Provide upstream and downstream process information

- Provide greater process integrity or disaster recovery

- Reduce costs

Notice that "cost savings" is last? This is not to say that a company should be extravagant, but people have become all-too carried away in recent years with cost savings. You could shut the operation down, then your costs would be zero, right? You obviously would not make any money for your company. The key is to select and fund technological platforms that support your most profitable core business operations. Spend money. Build that *other* side of the balance sheet instead—the income side!

4.3   What Is the Mission?

What is the one thing *every one* of your users has in common? The answer is, whether they are management or production, clerical or engineer, financial or entry level, *all* of them need to connect to *something*. (See Figure 4.1.) The technologist's role will increasingly be one of a business process facilitator and purveyor of "killer apps" which add to the organization's bottom line through technology. This adds a measure of both nobility and job security as the technologist's role becomes less of an "overhead" expense and more of a strategic contributor to the bottom line.

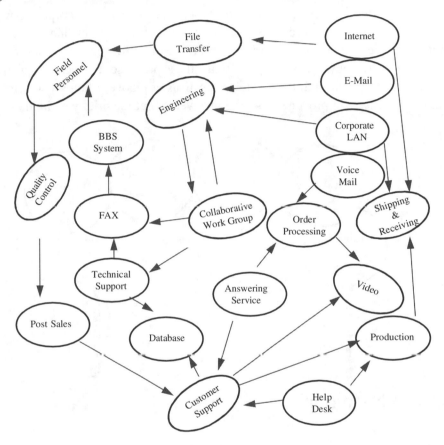

Figure 4.1: Everybody wants to connect!!! But first, what is the mission?

4.4   Examples of Killer Apps

Think about the financial success stories of the past decade. How about 1-800-FLOWERS? That is their name and number. Now forget the number. How's that for building customer loyalty? Think about that American Airlines Sabre Network mentioned earlier. What would airline ticketing be like today without it? Can you imagine pitching the business plan for that network in 1980? You might even have been accused of being a little crazy. But those crazy people are laughing all the way to the bank. How about the *Home Shopping Network*?

There are thousands of others. Can we learn anything from these examples? One thing is apparent, none of these companies used "Buck Rogers" technology. In fact many are based simply on intelligent use of communications to strengthen the customer support process and make products easy to order.

Recently, while performing a vulnerability analysis as part of a corporate-wide disaster recovery plan, we were forced to evaluate which parts of the business make the most contribution to the organization in order to prioritize them for recovery after a disaster. A very similar process holds true when prioritizing these processes for technological improvements. In fact, going through this exercise for disaster recovery helps you prioritize user applications for future improvements—and vice versa. Let me briefly explain.

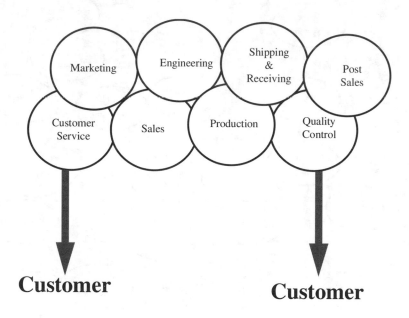

Focus First: Understand the big picture, then the rest becomes easier!..........

Figure 4.2: The mission is what best supports your customer.

The first thing you will note is that the process, like any good process, begins and ends with the customer. Starting on the left-hand side of Figure 4.2, you see a customer dialing in, probably on an incoming '800' number to a customer service center. Telecommunications service turns on the whole process, making it a revenue generator, since the business has no storefront.

Let's assume that the Customer Service Center has answered a call inquiring about one of the company's products, an electric motor. The customer works for one of the "Big 3" auto makers and is inquiring as to whether the motor is suitable for an automotive application. This is a tough question for the Customer Service representative, who is nontechnical. The call could however, represent a new market for the company. Therefore, rather than telling the customer, "Leave your name and number with me, and I'll call you when I get the answer," the Customer Service Center can respond in a very different way by virtue of its automated system.

While on the line with the caller, the Customer Service rep makes an inquiry *via data communications* to the LAN in Sales, to find out if similar sales for these types of automotive applications have been recently made. Perhaps the representative adds someone in Sales to the call. Sales, or Customer Service, in turn, can make an inquiry to the Engineering department LAN, illustrated by the third circle, as to whether it is feasible to adapt an existing product. Customer Service, Sales, and Engineering may make an inquiry to the Production people to learn how long it might take, for example, to adapt this product. You get the picture. By having fast access to information, the Customer Service Center can respond quickly, efficiently, and accurately to a customer's inquiry, *while the customer is on the phone and hot to buy*. What would have been *"I'll call you later,"* now becomes, *"Yes, we can adapt an existing motor line to your application. Engineering says it will take a week, and we can produce it within two weeks. How many would you like to order?"* You can just hear the cash register ringing!

The seamless integration of all of these disparate operations into one synergistic entity is probably the single, most important reason for migrating to a client-server architecture. By the way, anyone who says they are going to a client-server architecture solely to save money has, excuse the phrase, rocks in their head. Client-server is more expensive in the long run than a traditional centralized environment. But look at how well it works in empowering people and generating sales! This is the true value of a client-server architecture, to enhance that *other* side of the balance sheet we spoke about!

4.5   The Key to Success—The Network

Two things really sell with today's beleaguered end users:

- I can *manage* this better than you.

- I can provide you with more than you can get on your own.

Consider voice communications. We have essentially won that battle! With major corporations enjoying company-wide long distance rates in the low single digits per minute, few at the departmental or branch office level even consider buying their own long distance. How can we as technologists make data like voice? What does today's demanding data user really need? The answer is capacity, speed, and the ability to effectively manage and troubleshoot all components of modern Fast LANs, SONET, and ATM networks. Other "hot buttons" which can turn on your users are shown in Figure 4.3.

In short, when one considers the business criteria, it becomes apparent that capacity and effective network management may be the first catalysts your organization needs to turn a disorganized gaggle of distributed users into a comprehensive business solution. This book will cover strategies designed to make your LAN and WAN the principal catalysts behind this transformation.

CONNECTIVITY AND NETWORK MANAGEMENT ARE THE KEYS....

Reduce Expense   **Improve Reliability**   **Enhance Security**

Better Help Desk Support   **Better Telco Support**

Less People Intensive More Focus on Core Business   **Greater Efficiency**

SO WHY NOT USE YOUR WIDE AREA NETWORK AND NETWORK MANAGEMENT SYSTEM AS THE INTEGRATING FACTOR!

THIS BOOK WILL DESCRIBE USEFUL TECHNIQUES FOR DOING JUST THIS.

Figure 4.3: Keys to success.

4.6   Vendor-Managed Relationship

What if you are the vendor? The corollary to the "seamless solution" is called a "vendor-managed relationship." Think about all the things your company could do to make itself irresistible to its customers. Figure 4.4 should give you some ideas.

I once had the occasion to work with a specialty films company. This company did not make photographic film, rather, it made bags for Fritos Corn Chips, or dog food, or whatever. Another of their products was meat wrap, not a terribly exciting product.

The first thing the company did was make what I consider a courageous decision. They made the statement that their customers, pardon the phrase, would prostitute themselves to whomever had the lowest price on a given day. That was not a bad reflection on them, simply an acknowledgement of market dynamics. The price of meat wrap does not vary widely among suppliers, and everyone carries it. Even the "custom" Fritos bags could be handled by numerous suppliers. So what's a company to do? Apply enabling technology, that's what.

First the company installed special telephone numbers that pulled up account information even before the line was answered by screening the ANI (automatic number identifier). Next they linked the historical information in the database to items the caller had ordered before. For example, this gave the customer service agent the ability to see that the caller had ordered a special Thanksgiving bag for his corn chips the year before. If the call was to come in on October 1, the system would prompt the agent to ask if the caller wanted to order the "special Thanksgiving bag with the turkeys across the top" that they had ordered last year. It might even display a bitmapped picture so that the agent could describe the bag, or recommend changes. And you know what? Subtle prompting from a knowledgeable agent did in fact raise sales—significantly.

What about the meat wrap? Although it was difficult to glamorize this product, the company did install EDI (electronic data interchange) to major customers to allow ease of ordering. It started accepting orders over the Internet as well. This allowed its major customers to decrease inventory since they could order in a JIT (just in time) fashion and be sure of prompt delivery. The result? This company could offer a cost-saving JIT methodology to its customers, unlike its competitors. In this fashion they also installed themselves as a critical piece of their customer's production system and built great loyalty (and profit) for their trouble.

So you see? Even if all you sell is meat wrap, technology gives you the opportunity to build a "killer app" if you really understand your customer, and put your mind to work on how to meet his needs.

The other benefit to this applications-oriented approach is that your boss will *understand it*. This makes that presentation you ultimately need to make for funding go much easier. Tips for success follow in Figures 4.5 and 4.6.

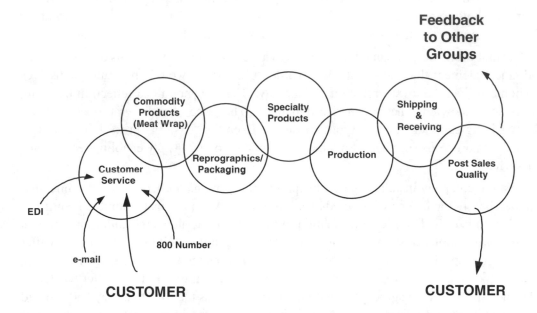

Before moving on, consider the techniques we showed you earlier, but this time set in the context of a Vendor-Managed Relationship. What kinds of interplay between customers and departments would you expect to deploy if you are a purveyor, rather than a consumer of services?

Figure 4.4: Vendor-managed relationship.

✓ Acquire good presentation skills, or get another individual to "champion" the project for you.

✓ Be upbeat and confident.

✓ Cover time frames, estimate of costs, and probability of success.

✓ Use BUSINESS and not TECHNICAL terms.

✓ Do your homework first.

✓ Go in "Armed for Bear."

Figure 4.5: The management presentation is important.

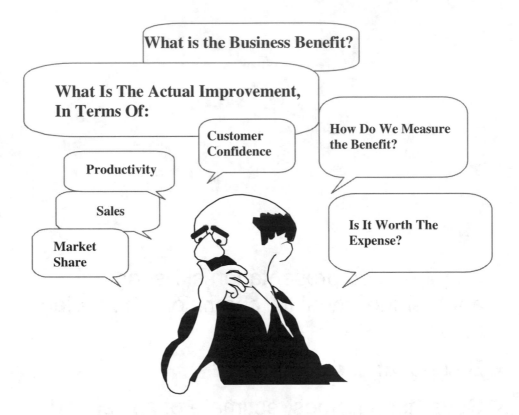

If this is explained in terms understandable to management, you will get support and funding!

The key is to build a case to which they cannot say NO.

Figure 4.6: Management needs the right information to make a decision.

4.7   Why Is Network Management Important?

As we said earlier in this text, two things are important (and sell well) to internal and external users. They are principally:

> "I can manage this better than you."
> and
> "I can provide you with more than you can get on your own."

From the standpoint of management of the environment, as well as from the perspective of classical economies of scale, the procurement of a superior network management system on behalf of your users qualifies on both counts above. It should be a major point of focus for any large Information Technology (IT) organization.

Any "seamless solution" or "vendor-managed relationship" is difficult at best or even risky from a disaster recovery standpoint unless thoughtful preparation goes into the systems that manage it for you. A well thought out network management system also helps you measure the results of your success for management.

4.8   Choosing the Right Network Management System

So what should be the evaluation characteristics of an effective network management system? The traditional evaluation criteria include capabilities in the areas of:

- Fault Management

- Operations Management

- Performance Management

- Configuration Management

- Change Management

- Security Management

- Accounting Management

Remember however, we do not just want this system to make our lives easier. We want to be able to show a concrete payback to the company by virtue of its implementation. This can be measured in "hard" dollars, such as reduced circuit costs, by virtue of being able to tune the network more efficiently. It can also be measured in terms of "soft" dollars, such as in greater customer satisfaction, by virtue of proactive maintenance capabilities, or better performance reporting. For example, if the system provides performance reports by circuit, or by circuit drop, these can be used by a sales staff or technical analyst to chart performance *by user*. The system might be a most effective power tool for you if used correctly and proactively.

4.9 Selecting the Right System

To be really effective in meeting user demand for network management services, as well as other core business objectives, a network management system should also have the following characteristics:

- The ability to interface with potentially dozens of distinct equipment outputs. Eventually the system will not only encompass the LAN, but also mainframe, data communications, and PBX systems. The more of these systems you can represent on a single display screen in your network control center, the greater the efficiency of your management system. Contrast this to shifting from screen to screen—or oftentimes from room to room—in present operating environments.

- The ability to satisfy the network management requirements of the user who is "black boxing himself to death." Your department can win these users back if it can manage the network better than the user can for himself.

- The functionality to be *proactive*, rather than *reactive* in nature. It should be able to read network data and take action *before* a problem becomes critical.

- The system should be intuitive and easy to use. Specific events should trigger plain English error messages. Rather than something cryptic, plain English scripts can often be written to banner "The Boston T1 circuit is about to go down!" This can be particularly useful if it happens at midnight on a Saturday night, when only one college intern is in the control center.

- The selected system should be inherently adaptable, even to the most proprietary systems, since you do not want to limit its capabilities. You often have no way of knowing in advance what platform a potential customer will be running. The more adaptable your system is across a broad cross section of products and inputs, the better.

- The system should sort and prioritize specific events proactively for long term analysis, network tuning, and optimization.

- It should be able to support specific network performance data *by customer*, for use in closer customer support by sales and account teams.

4.10 Packaging and Selling Network Management

Network management is extremely important to the overall operation of a technical organization. However, most end users can be expected to ignore the actual implementation of such a service insofar as installing it on their own. They will perform basic fire fighting on network problems, but generally lack the skills and tools to perform the other aspects of Integrated Network Management. *Consequently, an ongoing opportunity exists to provide these higher level network management services, win back end users, and increase organizational efficiency.*

Your organization should endeavor to provide a level of network management that is materially superior to what is available to your end user on his own. If successful in this regard, expenditures on a first rate network management system will prove not only a worthwhile investment, but one which will play a major role in your company's future success.

4.11 Defining the Components of a Network Management System

The complexity and distributed nature of networks used in today's *seamless solutions* demands more proactive management techniques and tools for successful deployment of seamless technology solutions. This is especially true for large companies, given their diversity of customer base and wide variety of supported services and platforms.

In addition, the pronounced shift into distributed systems will require an aggressive approach to management, since remotely located systems offer unique challenges. Those workstations which prompt the agent to "Ask the caller about the Thanksgiving Fritos bag he bought last year" add a significant level of complexity to the enterprise which is difficult to effectively control and manage. How many surplus CPU cycles exist in your organization for example right at this instant? In the mainframe world we could measure things like these. In the distributed world we probably cannot. How does one therefore know when they are about to run out of capacity? Without a good management system, you simply don't until something craters.

Numerous other limitations exist in the distributed processing community. For example, few of the standards which will govern these user-controlled systems exist today, even though these systems have immediate revenue impact and the effects of failure in them are pronounced. In addition, proprietary solutions to network management rather than industry standards are the norm. These systems work well in the *manufacturer's* environment, but most networks use a mix of vendor products. This creates numerous network management problems, not the least of which is having to bounce from one CRT to another to look at individual, proprietary systems, rather than having them all displayed on one screen, preferably in plain English.

Section 4.12 will attempt to crack the network management nut by clarifying the issues, and posing a few possible paths toward a solution. First, let's start with why we need a network management system.

4.12   The Purpose of a Network Management System

In a nutshell, the network management system is the eyes and ears of the caretakers of the network. As a company enhances its mainframe capabilities and develops its client-server–based business lines, the goal is to continue satisfying user demands for maximum network up time with minimal disruptions.

The performance of the network must be measured in terms of availability and quality that far surpasses what the customer can do for himself.

It is no secret that if the customer perceives he could do equally well or better without an information services or telecommunications department, the company loses a major incentive in producing new economies of scale. In the words of one internal I.T. executive, "Without a reliable network, what are we selling the user?" Therefore, foresighted companies must select the best available network management system that offers the best track record—and access points into the multivendor products that will inevitably exist on the customers' networks. This means that the network management system must outperform other vendor offerings, and do so by a wide margin.

4.13   Managing Network Diversity and Change

One major challenge technical managers face today is controlling their rapidly growing network environment. They feel pressure from business leaders to increase the effectiveness of their network and systems, while at the same time to reduce costs. Technical management is faced with incorporating new client-server technology into what has been a traditionally stable mainframe environment, while maintaining the same or better level of services to the end user. As one would expect, these newer technologies do not always integrate well into the existing infrastructure. Ultimately, the manager needs to create an environment that is open for information interchange and rich with interoperable solutions. The problem is that too few standards are available to make this an easily attainable goal.

In a multivendor environment, networks use multiple protocols and hardware platforms, often 25 or more distinct ones. This creates a problem when attempting to provide a truly seamless network management system. The lack of skill sets in the industry also adds to the problem, since many of the network management technicians and vendors have a propensity to be single platform or vendor specific. These problems are moving closer to the desktop, since users can add software or hardware locally. In fact, true network management encompasses so many areas these days that the options available for such a system can be limited.

4.14   Desirable Components of a Network Management System

The components required of a successful network management system include elements that allow systems and applications to be predictively controlled and managed. These include the following considerations:

4.14.1   Fault Management

Handling problems as they occur, or preferably before they occur, is a prime function. Most network management systems include at least a rudimentary level of fault management to isolate problems and alert network managers. Alarms and traps of various types are utilized, which specify predetermined thresholds so that proactive management can begin before the faults become critical. These can then be translated to common English messages for a network operator, such as, "THE BOSTON T1 IS ABOUT TO GO DOWN," for true predictive maintenance, especially on those Saturday nights when only a college intern is at the console.

Alas, potential vendors are working largely with operating systems and architecture proprietary to their specific product. Little has been done to fully integrate these systems into a single network management system. The vendor is more than willing to sell *their* proprietary network management tools but this would create a veritable potpourri of tools and terminal devices to perform this single function.

4.14.2   Operations Management

Capacity planning, optimization, and control of workload are all pieces that go hand in hand with the operation of a network. However, with emerging *seamless solution* architecture and the need to spawn remote procedure calls across the network, these pieces become increasingly important. The new system must reflect the new dynamics presented by a client-server environment.

4.14.3   Performance Management

Monitoring performance, problem recognition, data gathering, and evaluating information are standard parts of all network management systems. However, these components tend to be more reactive than proactive. The actual network management system should be intuitive enough to alert the network manager of performance bottlenecks, and if possible, begin a procedure for either adding capacity (if available) or shedding noncritical load. In the emerging client-server architecture, this one component may well be the foundation for the enhancements and capabilities of the implementations.

4.14.4 Configuration Management

The design and interworking relationships of all components on the network is critical to keep the network up and running. As the multivendor and multioperating systems are deployed within the client-server arena in support of *seamless solutions*, the need to control the interoperabilities is key to the success of the performance and stability of the network.

4.14.5 Change Management

Managing and controlling changes to hardware, software, updates, and new releases should also be a component that will assist in managing a network. As users or network administrators add components such as new software releases or hardware upgrades, the ever-present danger exists that things will deteriorate significantly. Hardware revision levels must be checked for compatibility to prevent unnecessarily bringing a network to its knees. The ability to recognize this change and troubleshoot the problem from remote management locations is essential.

4.14.6 Security Management

Control of access to a network must be handled proactively. In a network management system the access methodology, the considerations of what to do if a penetration is attempted, and the means of intercepting or disconnecting an unauthorized intruder on the network is required. When an attempt is made to penetrate the network, the network management system should be designed to immediately notify Operations of the situation. Internet access, and the new kinds of threats posed by it, come immediately to mind, as well as the ability to interact with the necessary firewalls.

4.14.7 Accounting Management

Gathering statistics and usage information for accountabilities or charge out purposes includes statistical information available from periodic and ad hoc reports. The accountability goes beyond gathering facts for charge out, and allows the network manager to gather historical and current statistics to predict the needs of the future. This allows the network manager to add capacity before the network becomes totally congested. In the customer care environments, these statistics can also help in responding to inquiries by account managers on specific performance data, by customer, for example.

4.14.8 Availability of a "Shrink-Wrapped" Solution

No off-the-shelf systems exist that meet *all* the demands of the network management system for *every* user. However, a number of standards have emerged that will offer some relief for the immediate future. As more time and development effort are expended by the industry on standardization, the odds of success rise accordingly. Functionally we can break larger problems down into more manageable smaller problems with a solid network management system.

4.15 Summary

Network management is extremely important to the overall operation of an organization's needs. However, outside of the corporate I.S. department, most users can be expected to ignore the actual implementation of such a service insofar as installing it on their own. They will perform basic fire fighting on network problems, but generally lack the skills and tools to perform the other aspects of Integrated Network Management. Consequently, an ongoing market opportunity exists for I.T. departments to provide these higher level network management services.

Potential prospects will therefore include internal users of large companies, or in the case of carriers and service providers, outside companies who do not implement a "Johnson Space Center" level of network management on their own for the following reasons:

- Lack of skill sets

- Lack of management support

- Inaccessible information

- Multivendor, multiplatform network

- Incompatibilities

- Nonstandard service offerings

- Just not enough time

To the extent that service providers and I.T. departments can fill this void and provide a level of network management that is materially superior to what is commonly available, expenditures on a first rate network management system will prove not only a worthwhile investment, but one which will play a major role in the provider's future success.

Note: Our sincere thanks to R. "Bud" Bates, President of TCI Consulting, Phoenix, AZ. for his kind contributions to this section. See www.tcic.com for more information.

5

High Speed WAN Connectivity (T1, T3, SONET)

5.1   Buying the "Big Stuff"

In many ways it is the large telecommunications user who benefits most in network procurements. Think about voice communications for example. Almost any large user commands certain economies of scales when buying voice services. There was a time, for example, when anyone who could send a million minutes or more per month to a long distance carrier could generally swing a "custom network deal" promising lucrative rates and terms. If anything, this has become easier as competition has grown more prolific. Even small companies can now swing single digit per minute rates on domestic long distance.

Even regulated long distance carriers, like AT&T, have offered the famous "Tariff 12" to their select users over the last 12 years. The terms of many of these "custom deals" can be reviewed by anyone inclined to pull the filings from the FCC, but few companies go through the trouble.

Bottom line, long distance is cheap, and voice is especially so. That brings two questions to mind:

1. How can we get the same "custom deals" on data?

2. How can we extend these "custom deals" to the local, as well as the long distance arenas?

We will touch on the technical side of these questions in this chapter by:

1. Showing you ways to shun "standard" telecom offerings, which were originally designed for voice, and poorly adapted later to data.

2. How to spec out and afford the "big stuff" in the form of "native LAN" connections.

3. How to make your data network like voice, in terms of standardization, ubiquity, cost, and ease of use.

4. How to position for the future of voice and data technology now

5.2   The "Old Stuff"—T1

T1 is not new. Pulse code modulation technology for T1 was actually conceptualized in the late 1940s and saw widespread deployment in the public network in the early 1960s. When users began demanding larger pipes for data transmission and for other uses, T1 was the logical choice since at the time, this was the biggest thing that the telcos could easily offer.

T1 was not designed to handle data. In fact, T1 was designed to run over copper cable to primarily carry voice, but was adapted by users and telcos to carry data. Everyone adapted their equipment to run over T1 since that is all the telephone companies had. I mean, really, can you imagine a data processing manager voluntarily running his 10 or 16 Mbps LAN interconnections at 1.544 Mbps by choice?

In its most elementary manifestation, shown in Figure 5.1, a channel bank was installed on the customer's premises, along with a matching channel bank in the central office. Twenty-four voice or data channels were carried over the T1 circuit. However, there were a few problems with this. First off, the technology is dumb. If things start going haywire, the only thing that can usually be done is to take the whole channel bank down to troubleshoot, usually putting at least 24 users out of business. We know the likely outcome of this action. Second, since the original channel banks were designed to carry voice, they robbed portions of the T1 for signaling information. This meant we did not get the full 1.544 Mbps for our data. In fact, we did not usually even come close. We could go on, but suffice it to say we have spent 20 years being shoehorned into a technology never originally intended for data. T1 was designed for interoffice toll telephone circuits meant to talk over, plain and simple.

Users eventually got around some of these problems by installing proprietary multiplex equipment on either end of the T1 at each customer location. The telephone companies made the first attempts at providing nonintrusive maintenance, namely, ESF or extended super frame capabilities. Users continued to develop applications that outpaced the capabilities of the telephone company by every measure—size, speed, network management, reliability.

Consider Figures 5.1 and 5.2. While they both appear a little dated, most of the networks in place today use the same basic technology as shown in these illustrations.

5.3   How We Have Done Business

Table 5.1 shows the traditional U.S. and International digital hierarchies. Note in particular, that the only way to link a North American T1 (24 channels) with a European E1 (30 channels) was to drop to the DS-O (single channel) level. This meant taking a nice clean digital signal, converting it to analog (individual channels), only to mux it up again to digital. This is one of the reasons (only one) that overseas calls sounded so bad until just recently.

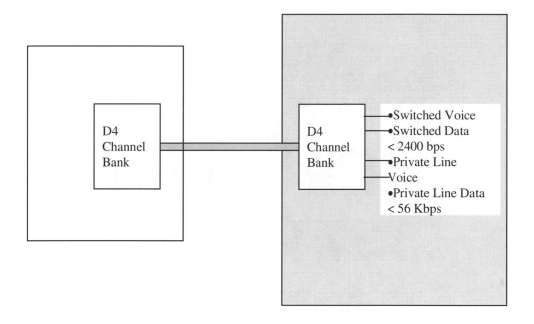

Characteristics:

- •24 individual channels
- •Speed of 1.544 Mbps
- •Designed originally for voice but adapted to data
- •Maximum data speed 56 Kbps

Figure 5.1: Traditional T1 service (circa 1985).

Level	Europe	Japan	North America	Transatlantic
0	64	64	64	64
1	2.048	1.544	1.544	2.048
2	8.448	6.312	6.312	6.312
3	34.638	32.064	44.736	44.736
4	139.264	97.728	139.264	139.264

Table 5.1: U.S./International digital hierarchies.

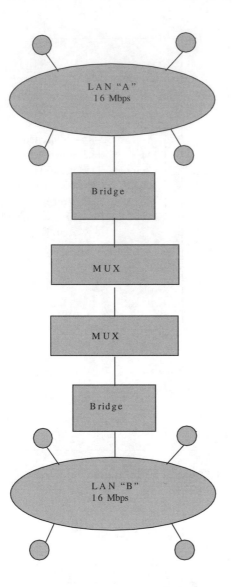

Telco T1 circuit 1.536 Mbps
(at best) or approximately
10% as fast.

Most LAN users can relate to the experience of "flying along" on a LAN at 16 Mbps, only to be "throttled down" to the speed of 1.5 Mbps when interconnecting to another LAN. This is one classical case of the limitations we had on networks in the past.

Figure 5.2: Traditional LAN interconnection bottleneck.

5.4   Satisfying Today's Users with Tomorrow's Technology

For many of us, the day has arrived when we can call the telephone company and say, "I need a 16 Mbps circuit to Boston" with a completely straight face. The tools also exist for a Johnson Space Center level of network management, including the ability to call your carrier and say "Hey, the Boston 16 Mbps circuit is *about* to go down!" If you have a user who needs a 114.53 Mbps data circuit (and has a darn good reason for needing it) you might just be able to provide it to him now. And most importantly, it is possible now to provide all the ergonomics of the personal computer at the desktop—without the moving parts (and attendant operation costs) of the PC. Connectivity, Capacity, and Network Management—these are the keys which will enable tomorrow's technical services manager to unite a disorganized gaggle of distributed users into a comprehensive business solution. The time to start is now, with the advent of an enabling technology called SONET (Synchronous Optical Network), which is discussed below, and in greater detail at the end of this chapter

5.5   What Do Large Users Want?

What are the two most common "hooks" the network manager has in meeting goals like those above? They can be basically boiled down to two statements:

> *"I can manage this better than you."*
> *"I can provide you with more capacity than you can get on your own."*

The network manager can make good on these two pledges through the use of radically advanced network technologies becoming available right now. Frame Relay, SMDS, and ATM are always bantered about in technical discussions. However, the point of commonality for all of these technologies is a new, improved transport network technology called SONET, or Synchronous Optical Network.

5.6   What Is Great About SONET?

For starters, you can recommend SONET and generally not screw up! Here is why. In addition to changing the dynamics of everything, SONET will be a 30 year technology (and how often do we get *that* long a planning window!). It is destined to replace T1 and T3 as the backbone transmission media in North America and worldwide.

The wide area network is getting smart. SONET is almost unbelievably smart by today's standards. For example, much of the work we do for disaster recovery planning and network diversity is becoming an integral component of the network itself. This is not future stuff, it is here today.

Your future network management system will need to read, interpret, and act upon the data provided to it by these new intelligent network capabilities. Since everything except the glass in a SONET network is smart, every component will be reporting in regarding its condition and performance. Other capabilities, such as the ability for a SONET user in Minneapolis to "drop" a circuit off at a node in Dallas without ever leaving his armchair, bode well for shortened circuit provisioning times. The day is rapidly approaching when a user will be able to call a carrier and say "I need a 10 Mbps circuit to Atlanta" to which the carrier replies almost immediately, "Do you see it *now*?" Circuit activation to users with high capacity access will become as easy as turning on a new cable TV channel.

But first, some background. In order to truly appreciate these changes, let's look first at today's mainline transport system in the US, the T1.

5.7 "The Old Days"

Wow! What a difference 15 years makes. Remember the days when a coast-to-coast call *sounded* like a coast-to-coast call? By anyone's standards, the quality difference between the analog transmission of yesteryear and today's digital capacity is like comparing 78 rpm records to compact discs.

5.8 Today's Technology

The word for today is *digital*. Whether it is an electrical signal like T1, or a photonic signal like fiber optics, digital carriers regenerate the human voice every few miles, rather than first amplifying voices (and noise) like the old ones did. That is why today, London, England sounds more like London, Ontario.

5.9 What Is State of the Art?

That is easy. As far as carrier systems go, SONET represents the state of the art. It is digital, it is *smart*, and it is discussed in detail at the end of this chapter.

5.10 What Is Out There Now?

Over 70 percent of networks employ more primitive AMI and M13 technologies. This chapter will discuss the differences, and make a case as to why "Fortune" users need SONET.

5.11   How Do We Position? What Is the Migration Path?

That is the 64 Kbps question! Significant differences (and shortcomings) in high-end carrier technology, what they mean to "Fortune" users, and how to chart a migration path to the network technologies of the next millennium, will be shown throughout this chapter.

5.12   What Is a T1?

A T1, colloquially speaking, is a 1.544 Mbps data circuit. It is used to take 24 voice channels and multiplex them using something called PCM (pulse code modulation) into a digital signal that can be carried on four wires. It replaced interoffice trunks (circuits between Bell central office switches) years ago. T1 has been around for 30 or 40 years, but only since around 1980 has it been installed at customer locations. One important point to consider: even though large corporate users use thousand of T1s nationwide, in many cases, they use T1 for a purpose for which it was never intended—carrying data. T1 was designed for voice, not data.

What if, however, you are using an older protocol like AMI (alternate mark inversion) which steals the eighth bit in each channel for timing? In that case you have only 7 bits left, multiplied by 8,000 bps, or a 56 Kbps circuit.

By now, you may be totally lost, but I wanted you to see how engineers come up with the different line speeds we toss around. You have undoubtedly heard of a 64 Kbps circuit or a 56 Kbps circuit. The difference is the protocol.

That eighth bit can also be robbed by the telco to switch the circuit. In fact, if you made an interexchange phone call today, you really spoke over a switched 56 Kbps circuit. Did you hear the difference? I think not. Unless you are Lassie or Rin-Tin-Tin, the call sounded fine. But data managers notice! 56 Kbps × 24 channels is only 1,344 Mbps—a lot less than the 1.544 Mbps the data manager is paying for. (See Figure 5.3.) That is why engineers came up with a more modern line code called B8ZS (Bipolar 8 with zero substitution).

Diagrams of the T1 frame appear below (see Figures 5.3 and 5.4). They are not as complicated as they first appear. Think about it. The sampling rate of a T1 is 8,000 times per second. Each sample is 8 bits. That is 8 × 8000 or 64,000 bits per second. There is your 64 Kbps channel! There are 24 channels in a T1; 24 × 8 = 192 bits. Add one for framing and you get the 193-bit T1 frame illustrated. You can look at each 8 bit "time slot" and imagine just where channel 1, channel 2, channel 3, and so on, are. It should also be apparent from looking at the diagram that you do not need to use channels at all. You can use all 192 bits as one big pipe for data, just like many companies do.

Framing Bit Requirement
193

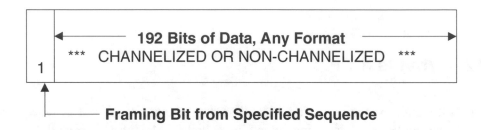

192 Bits of Data, Any Format
*** CHANNELIZED OR NON-CHANNELIZED ***

1

Framing Bit from Specified Sequence

- **D1 (Single-Frame Format)**
- **D1D, D2, D3, D4 (12-Frame Superframe)**
- **ESF (24-Frame Extended Superframe Format)**

Figure 5.3: T1 frame.

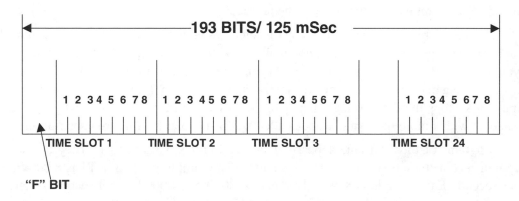

193 BITS/ 125 mSec

1 2 3 4 5 6 7 8   1 2 3 4 5 6 7 8   1 2 3 4 5 6 7 8   1 2 3 4 5 6 7 8

TIME SLOT 1   TIME SLOT 2   TIME SLOT 3   TIME SLOT 24

"F" BIT

Figure 5.4: T1 framing overview.

5.13   T1 Protocols

Like we have already stated, T1 has always been a 193 bit frame. Of these 193 bits, 192 are for channels. A T1 "samples" digital signals 8,000 times per second and uses an 8 bit "word" as it samples. Eight bits $\times$ 8,000 times per second = 64,000 bits per second, or a 64 Kbps channel. There are 24 64-Kbps channels in a T1. Twenty-four $\times$ 64,000 = 1.536 Mbps or one T1. Lastly, there is overhead in a T1. Every 193rd bit is used to frame the signal. One $\times$ 8,000 = 8 Kbps. So, 1,536 + 8 Kbps = 1,544 Mbps. All bits present and accounted for!

5.14   B8ZS

If you have ever ordered a T1, one of the first things the telco usually asks you is whether you want AMI or B8ZS. The old protocol we discussed, AMI, was a bit robbing protocol. That means it takes the *eighth* bit of each channel (look back at Figure 5.4) to do things like frame the T1 and make sure it does not lose track of where the channels are. The eighth bit can also be used to carry signaling information which is discussed later. Whatever way you cut it, each bit represents 12 1/2 percent of your T1 capacity so you do not like to give them away if you do not have to. That is why they invented B8ZS.

B8ZS is a bit *substitution* protocol. Rather than steal bits from you, it substitutes a "secret code" or unique sequence of ones and zeros which maintains timing on the T1, but does so at no cost to you.

The problem with data on a T1 is that data can generate long strings of zeros. When you send a long string of zeros over a T1, there are no ones in the bit stream to remind it of where it is. At 1.544 Mbps, it does not take long for it to lose its place. Then when the ones start up again, it is "out of synchronization" and flops. (Ever hear of a *frame slip* on your T1?) That is why the telco has a rule that 12 1/2 percent of the carrying capacity of a T1 *must* be ones. And the most convenient and consistent way to always meet that rule is to steal every eighth bit to keep your place.

The telco will ask you whether you want AMI or B8ZS when you order a T1. If you use a "bit robbing" protocol like AMI, you only get seven bits per channel. Seven $\times$ 8,000 = 56,000 or 56 Kbps. If you use a "bit substitution" protocol like B8ZS you get all eight bits. Eight $\times$ 8,000 = 64,000 or 64 Kbps per channel. Based on these facts, which do you want if you are ordering a T1 for data? For voice it does not matter, only dogs can hear the difference. For data though, B8ZS is the ticket.

Here is the rub however. If your telco does not provide B8ZS, it is expensive to get it, since most all of their equipment needs to be replaced. If it is available, you will recoup 192 Kbps of bandwidth for your data. The top speed for a T1, with very few exceptions is 1.536 Mbps, not 1.544. That is because the telco needs that 193rd bit (1 Kbps) to frame the T1, and also provide ESF (extended super frame) diagnostics.

Figure 5.5 graphically illustrates how B8ZS works.

5.15   ESF

What is ESF? In essence, ESF provides you and the carrier with the ability to monitor and troubleshoot a T1—without taking it down! (See Figure 5.6.)

In the old days, a user's first sign of trouble was generally when the T1 went completely out of service. Moreover, T1 never checked for errors in content. Since it was designed for voice, it was simply assumed the caller would repeat his last phrase—that is, error correction in the voice world.

Data presented a different problem. If in a financial transfer, a "0" was changed to a "9," that presented a problem. ESF will not *correct* that kind of thing, but it will let you know when it happens. For those of you interested, ESF takes that 193rd bit and changes its meaning, depending on when it is received. For example, the bits come in this order:

DL (Data Link)
BC (Block Error Check)
DL (Data Link)
Fe (Frame)

The framing bit changes in meaning depending on when it hits: DL bit, BC bit, DL bit, or Frame. In the 24 frame superframe, the DL bit hits 12 times. The other two hit six times each. This actually builds these distinct and nonobtrusive functions into the T1, on a bit which was originally overhead anyway. The framing bit hits one time each frame, and the T1 samples 8,000 times per second. One bit $\times$ 8,000 = 8 Kbps. That 8 Kbps is effectively divied between the DL, BC, and FE bits (Data Link, Block Error Checking, and Framing bits), Since the DL bit hits every other time (50 percent) it is 4 Kbps. The other two hit every fourth time (25 percent of the framing bits) so they are 2 Kbps each. See Table 5.2.

Rather than use all the bits just to frame, we have now added some "smarts" to T1. The BC bit runs a CRC check to tell you when errors occur. The DL bits form a maintenance channel to constantly monitor, and even record, T1 status info. Imagine two T1 CSUs in constant dialogue every second, all day, every day:

"How do you feel?"
"I feel great! How do you feel?"
"I feel great too."

The BC bit takes the T1's temperature, so to speak, and the far end reports back using the data link channel. All this occurs in what used to be lost overhead used just for framing.

- ✓ **B8ZS line code**
- ✓ **CSU looks for 8 zeros**
- ✓ **CSU automatically substitutes**
- ✓ **00011011 over the line**
- ✓ **Distant CSU converts 00011011 back to the original 8 zeros**
- ✓ **Causes a BPV on frames 4 and 7**
- ✓ **Carrier must upgrade equipment**
- ✓ **Result: AMI = 1.344 Mbps**
- ✓ **B8ZS = 1.536 Mbps**
- ✓ **(Recoups 192 Kbps bandwidth)**

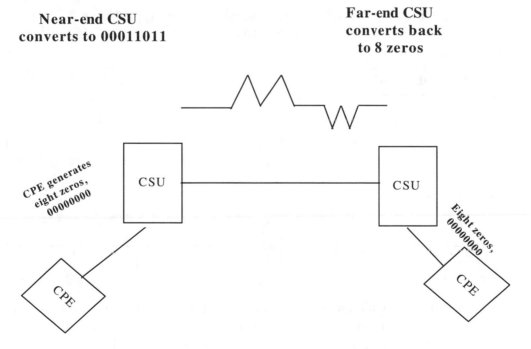

Figure 5.5: Bit substitution.

- One framing bit, 24 eight bit channels

- EXTENDED SUPER FRAME consists of 24 frames

- The ``S'' bit of each frame forms an 8 bit channel,
  further divided into three subchannels:

Fe   DL   BC

Fe = 2 Kb (Framing)

DL = 4 Kb (Data Link)

BC = 2 Kb (Block Error Checking)

**Thus, the ``lost'' 8 Kb normally used for framing performs 3
functions**

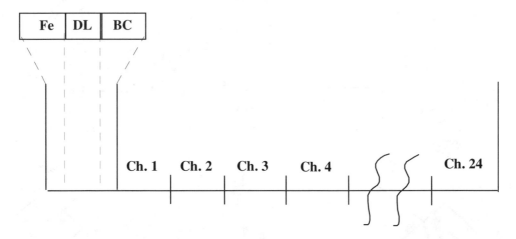

Figure 5.6: DS1/ESF framing format. Extended super frame makes use of the framing bit by changing its meaning over time. The 8 Kbps that used to be wasted only on framing now provides two other useful features—CRC error checking and a data link channel.

	S Bits		
Frame Number	Fe	DL	BC
1	–	m	–
2	–	–	C1
3	–	m	–
4	0	–	–
5	–	m	–
6	–	–	C2
7	–	m	–
8	0	–	–
9	–	m	–
10	–	–	C3
11	–	m	–
12	1	–	–
13	–	m	–
14	–	–	C4
15	–	m	–
16	0	–	–
17	–	m	–
18	–	–	C5
19	–	m	–
20	1	–	–
21	–	m	–
22	–	–	C6
23	–	m	–
24	1	–	–

Table 5.2: Extended super frame overview (assignment of S bit in DS-1 signals).

5.16   Fault Conditions

ESF monitors for out-of-frame conditions, bipolar violations, errored seconds, and other conditions. The difference between an ESF and a standard T1 is like night and day. When you order a T1, the telco will ask you if you want ESF. (AT&T *requires* it.) Based on this information, how will you respond?

5.17   T3

You probably deduced from the last section that if you entered an AMI T1 and a cinder block in an intelligence contest there would be a *tie*. That is a fair assessment. ESF made T1 a little smarter, but remember, it was never designed for data. ESF at best is a Band-Aid fix to the same old hog. T3, to put it bluntly, is worse.

Let's start with what T3 was designed to do. Way back when, before the 1984 AT&T divestiture, there were 10 AT&T "Class 1" offices. Each served up to a five state region. Needless to say, these guys generated some traffic.

T3 was designed to be nailed down one time, between AT&T Class 1 COs, and never move. Everything that is sacred to data managers was no big deal in the original T3 world. There was little in the way of diagnostics, and a single often minor alarm took down all 672 channels. No big deal to a carrier with dozens of T3s, but a career-threatening blunder to the data manager.

5.18   M13 Headaches and Shortcomings

As you can see from Figure 5.7, M13 is not like T1. As far as being understandable from a visual standpoint, it is a travesty. M13 uses a complex, nonlinear frame structure that just does not add up. For example, there are 4,760 bits per frame, and T1's 193 bits will not divide into this. The protocol is also different. Rather than B8ZS it uses B3ZS—no more than 3 zeros due to its faster speed.

There are one or two things which are recognizable. There are seven subframes per M-frame which correspond to seven T2s contained in the bit stream. For that reason an M13 is really an M1-2-3. Each subframe has eight 85 bit slots, and each slot has one bit for control, 84 for payload.

The two-step multiplexing process *bit interleaves* one bit at a time! Unfortunately when M13 equipment was designed, buffer memory was a little hard to come by. That is why each channel is mapped in one bit, not eight bits, at a time. That is also why DS0s are not identifiable or intact. Twenty-eight T1s map into an M13. Each of these T1s are effectively running at *different speeds* since the T1 spec is 1.544 Mbps *plus or minus 75 bps*. Since there is no buffer, the only thing the M13 can do is slow down the fast T1s. That means adding lots of overhead bits. That is why 28×1.544 does *not* = 44.736 Mbps because there is a lot of bit stuffing going on.

5.19 Ski Lift Analogy

Think of it another way. Imagine your company is one of 28 companies taking a company picnic at a ski lodge. Each of the 28 companies has 24 employees. All 28 companies form 28 lines for the ski lift. The problem is, there is only one lift. The idea is for each company to line up, each employee to go up the ski lift, ski down the slope, and then very importantly, get *back* in the same line to go to the hospitality tent for pizza and beer. Since your company does not want to buy pizza and beer for nonemployees, it is imperative that everyone gets back in the same line.

All of the lines are moving at different speeds, and that is creating problems. Soon the management sees what is going on, and realizing that everyone has been drinking all day, decides to add a few policemen to the line for crowd control. At the front of the 28 lines, there is a policeman who tells everyone, "Form seven lines!" (the seven T2s).

At the front of the seven lines there are more policemen, and finally everyone gets to go up the lift. But there is another problem. They do not go one employee at a time. *They go 1/8 of an employee at a time.* When you look at the lift, you can see Ralph's leg, and Karen's elbow, and Jack's neck and left foot, but you cannot really recognize any employee. Even the policemen are mixed in with the employees. It is only at the other end where they "reassemble" the employees, strip off the policemen, and get everyone back in the correct line again that anything becomes recognizable.

One of the problems with this is that if any employee wants to jump off the ski lift midstream—he cannot. He is all in pieces! That is why you cannot easily drop and insert channels, for example, with M13 T3s! And that is only where your problems begin!

Let's say a carrier or user in Boston installs a T1 to San Diego. This is straightforward enough. Let's say, however, that they want to drop one channel off in Denver. It takes almost 60 boxes to do that, since the entire T3 (using M13 technology) must be broken down.

The only thing that will break down an M13 is another M13. It also causes one of those nasty A to D conversions we spoke of earlier, and messes up the whole T3 for everyone. Part of the problem for M13 lies in its bit interleaved and archaic frame structure. Consider Table 5.3 entitled M13 frame. If you liked the T1 diagram, you ought to *love* this one. It really means nothing, except to another M13. The reasons behind this can torture graduate level engineers for days, but we will spare you for the purposes of this book.

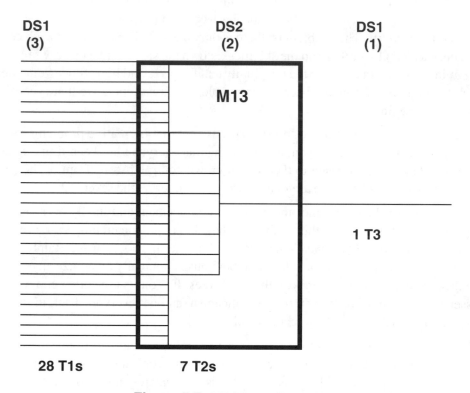

Figure 5.7: M13 is *really* M1-2-3.

5.20   The DS3 Signal

The DS3 signal is partitioned into frames of 4,760 time slots, divided into seven frames having 680 time slots: (4760/7=680).

The Subframe is further divided into eight blocks of 85 time slots, 84 for information, one as a control bit: (680/8=85).

The X control bits, located in the first and second subframes, are a low speed signaling channel.

The P control bits in the third and forth subframes, count parity and insert it in the following M frame.

The M control bit in the fifth, sixth, and seventh subframes creates a multiframe alignment signal which locates the seven subframes, while F control bits are used to identify all the control bit time slots.

C control bits are used in accordance with the DS3 protocol selected. This is the only protocol-dependent bit slot. The only difference in the major T3 protocols is how they interpret the C bits.

T3 protocols—As I said, the three major DS3 protocols differ only in their interpretation of the C bits, and the 85 bit information time slots. They include M13, C-Bit parity, SYNTRAN, and SONET.

C-Bit parity—In 1991 ANSI T1.107 made C-Bit an official standard protocol. AT&T originally developed C-Bit as a "field upgrade" to M13. That means M13 is being replaced by C-Bit.

C-Bit supports channelized and nonchannelized bit streams, meaning users of traditional T3s no longer have to spring for expensive inverse multiplexes (IMUX) to run video for example. Both C-Bit and M13 share M13 frame structure, but C-Bits are redefined as shown below.

C-Bit Definitions:

- 7, 8, and 9 carry path parity bits

- 10, 11, and 12 carry FEBE (Far End Block Error) checking end to end

- C-Bit 1 is a format ID

- The remaining 14 are presently undefined

5.21   C-Bit Parity

C-Bit parity sounds a lot like ESF, doesn't it? In a way it is. C-Bit is something that made M13 just slightly more intelligent. Like ESF, however, it too is a Band-Aid fix to the same old hog. C-Bit parity, like the name implies, redefined the C bits in a T3 to add diagnostic and error detection capabilities, just like ESF did for T1s. Tables 5.3 and 5.4 show a few of the new features for those of you inclined to look. Data links channels have been added, just like in ESF. Moreover, FEBE (Far End Block Error) checking and other features now ride transparently in the T3. One good thing about C-Bit was that it was relatively cheap. An M13 to C-Bit conversion was a "field upgrade." Today, much of what is considered "installed base" is C-Bit parity. Based on these facts, if your carrier gives you the choice between M13 and C-Bit, which is better?

	-85 slots-		-85 slots-		-85 slots-		-85 slots-		-85 slots-		-85 slots-		-85 slots-		-85 slots-	
1	X	[84]	F1	[84]	C1	[84]	F0	[84]	C2	[84]	F0	[84]	C3	[84]	F1	[84]
2	X	[84]	F1	[84]	C4	[84]	F0	[84]	C5	[84]	F0	[84]	C6	[84]	F1	[84]
3	P	[84]	F1	[84]	C7	[84]	F0	[84]	C8	[84]	F0	[84]	C9	[84]	F1	[84]
4	P	[84]	F1	[84]	C10	[84]	F0	[84]	C11	[84]	F0	[84]	C12	[84]	F1	[84]
5	M0	[84]	F1	[84]	C13	[84]	F0	[84]	C14	[84]	F0	[84]	C15	[84]	F1	[84]
6	M1	[84]	F1	[84]	C16	[84]	F0	[84]	C17	[84]	F0	[84]	C18	[84]	F1	[84]
7	M0	[84]	F1	[84]	C19	[84]	F0	[84]	C20	[84]	F0	[84]	C21	[84]	F1	[84]

Table 5.3: M13 frame.

M1	C-1 FID	C-2 N/U	C-3 FEA	FID & FEA
M2	C-4 DLa	C-5 DLa	C-6 DLa	Data Link a
M3	C-7 CP	C-8 CP	C-9 CP	C-Bit Parity
M4	C-10 FEBE	C-11 FEBE	C-12 FEBE	FEBE
M5	C-13 DLt	C-14 DLt	C-15 DLt	Data Link t
M6	C-16 DLi	C-17 DLi	C-18 DLi	Data Link i
M7	C-19 DLa	C-20 DLa	C-21 DLa	Data Link a

Table 5.4: C-Bit parity. Each C bit occurs 9,398+ times per second (9.4 Kbps).

5.22 SYNTRAN

We spoke of the C-Bit "Field Upgrade." This next option was a "Forklift Upgrade." SYN-TRAN, developed by Bellcore, solved most of the problems inherent in M13 by completely redefining the frame structure. One of the most important advancements was the introduction, for the first time, of an efficient means of dropping and inserting channels.

Unfortunately, it was too late. By about 1990, when SYNTRAN was just becoming established, our prayers were answered—SONET came along.

5.23   SONET!—A "Smart" Network

Enter SONET, the "smart" network technology. One of the first areas of impact by SONET networks will be in the area of network management. By integrating the capabilities of a state-of-the-art *network management system* with a feature-rich SONET transport system, the face of internetworking will be changed forever. SONET features and capabilities can be adapted and groomed to address pressing business issues, chiefly in terms of:

- Reduced expense

- Improved reliability

- Streamlined telephone company coordination and troubleshooting

- Higher bandwidth and capacity

We will illustrate how SONET will change the face of networking. For example, the distributed networks of today will quite possibly evolve into a sort of "centralized client-server." LANs will still exist to be sure, but they will be chiefly centered around "knowledge" based workers like attorneys, engineers, graphic artists, and others who need the functionality of a PC or require an oddball application. For repetitive tasks that are easily automated, a more centralized approach will make the best economic sense, and will be the easiest to support in terms of operations cost. The graphical display interfaces will reside at the work location, but the meat of the system will reside in a secure, centralized location using the horsepower of the SONET network to pump graphical and multimedia information to the end user. The advantages? Greater security, greater standardization, reduced equipment cost, and drastically reduced personnel cost at the customer location. The requirement to do this effectively? Communications capacity, lots of it, with an unparalleled ability to maintain and control it. That is one area where SONET comes in. It provides the capacity to install applications like these combined with the command and control and network management features up to par with mission critical production systems.

5.24   SONET Framed Signal (STS-1)

SONET cleans up a lot of problems. First, it maps in circuits eight bits at a time. It is *byte* interleaved instead of *bit* interleaved. The frame is 90 *bytes* wide by 9 *bytes* high for a total of 810 *bytes* or 6,490 bits. Okay, let's do a little math again. Eight thousand times per second multiplied by 6,480 bits equals 51.84 Mbps, the OC1 line rate. Great, that seems simple enough. No policemen or other overhead in the payload section.

The overhead section is 27 bytes, or 3 bytes by 9 bytes, all at the beginning. Everything else maps in very logically, one channel at a time, left to right and row by row. If you would like to find the seventh channel of the 13th T1, you know exactly where it is because, unlike M13, it remains intact. The overhead is also intelligent and contains things like protection switching, orderwires, and other functions (see Figure 5.8).

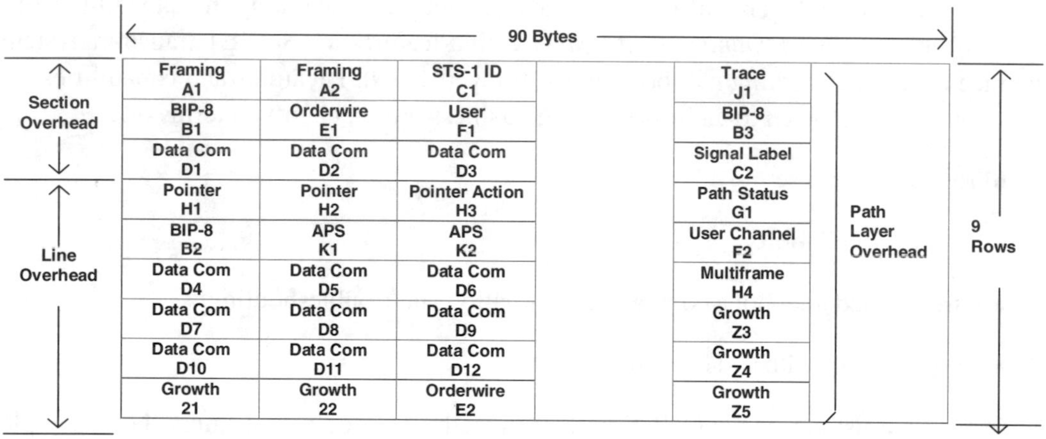

Figure 5.8: Overhead byte locations in an STS-1 frame. *Source:* ANSI.

5.25   SONET Link Architecture

Carriers refer to specific areas of a SONET network in terms of sections, lines, and paths, and corresponding equipment as STE, LTE, and PTE. Since many times these functions are all contained in the same box, the equipment is simply referred to by the designator "SONET PTE." That is where we got the phraseology used in this book. The sections, lines, and paths are illustrated in Figure 5.9.

5.26   Other SONET Points

5.26.1   SONET Is Inefficient—So What?

One seventh of the available bandwidth is available for overhead, and is scalable on demand Why not? It's fiber, waste the bandwidth!

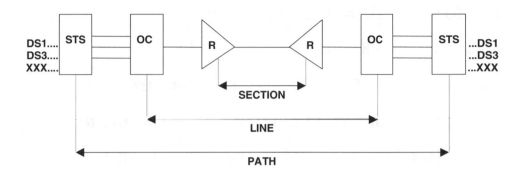

Figure 5.9: SONET components.

5.26.2 Plenty of Room for a Brain

51.84 Mbps − 44.736 Mbps = 7.1 Mbps. That overhead is the same for every OC1. An OC3 has 3 times 7.1 Mbps or 21 Mbps overhead. An OC12 has 12 times the overhead or around 84 Mbps. That seems inefficient but remember, fiber is like Doritos, eat all you want and we will make more.

5.26.3 Smarts in all the Boxes

Everything in a SONET network is smart except the glass. It has memory, processors, intelligence! Every box and repeater in the network is *smart*—much of what you once needed test equipment for is now done by the boxes! Bit error rate tests and other diagnostics can be accomplished by the equipment itself!

5.26.4 No Hassles with CRC Checking

Remember all the error checking we did with ESF and C-Bit? Forget it now. Fiber is clean and virtually impervious to impairments like RFI, EMI, and so forth. The only things that consistently bother fiber are backhoes and rats. Therefore, all error checking is done not in the SONET frame but at higher network levels.

5.26.5 Convenient Connection to SDH

SONET is essentially a North American standard. The rest of the planet uses Synchronous Digital Hierarchy or SDH. Even so, SONET connects to SDH at all the most important rates (see Figure 5.10).

- **STM1 = 155.52 = OC3**

- **STM4 = 622.08 = OC12**

- **STM8 = 1244.16 = OC24 (recently standardized)**

- **STM12 = 1866.24 = OC36 (recently standardized)**

- **STM16 = 2488.32 = OC48 (recently standardized)**

Even though they are more compatible, they are *still* not *fully* compatible in terms of framing, format, and speed.

Figure 5.10: What about the rest of the world? SONET does not hit every worldwide rate, but close enough.

5.27   What About the Fiber Itself?

5.27.1   Multimode

Multimode fiber has a comparatively wide *core* and is usually what you find in your building. It is generally good up to about 622 Mbps or the OC12 line rate. That is partially because it is powered by LEDs (light emitting diodes) and not lasers. It is also because the relatively large diameter core leaves ample room for light to bounce around and get out of phase.

5.27.2   Single Mode

Single mode fiber is what the telco uses. It has a narrower core and is powered by *lasers*. With single mode fiber, as far as bandwidth is concerned, the sky is the limit. Presently the theoretical limit for single mode fiber is some 4 *terabits* per second. And that is before some wise guy asks, "Why only use *white* light? How many *colors* are in the spectrum?" And that is just what is going on now with something called Wave Division Multiplexing. The next generation of WDM equipment (1999) will be OC3072 (16 times the OC192 rate which is currently state of the art)!

5.28   Summary

After reading this chapter you should understand the reasons why SONET is a technology that technologists can recommend and not screw up. Since it will replace T1 and T3 as the backbone transmission media, it will be the platform for ATM and other emerging technologies. It meets our criteria for network management, that is, "I can manage it better than you." It meets our criteria for greater bandwidth, that is, "I can provide you with more than you can get on your own." And even more importantly, it is available now and with the advent of local competition, all of the carriers want to get it into your facility. So what are you waiting for? Begin to negotiate your "Custom Network Deal" today with SONET access as its cornerstone. The rest of the book will explain how.

6

Advanced Techniques for High Speed WAN Connectivity

6.1   "Buck Rogers" Telecom Applications

Look into your crystal ball to the future. The year is 2020. Let's talk technology. It may sound preposterous to try and second-guess technology 20 years into the future, but in a few specific areas, such prophesying is not only possible, but advisable. Many of today's technologies are doomed to an early death and in fact will be obsolete within two or three years. Personal computers are a good example, since it is a well-known fact that technology changes at least every 18 months. But what about telecommunications planning? Many telecommunications technologies will also be obsolete within the same period, to be sure. However, a few technologies, namely the *transport technologies*, will enjoy a long service life. One example of these technologies is SONET or Synchronous Optical Network. See Figure 6.1.

SONET is the technology that will replace T1 and T3 as the backbone transmission media throughout the United States and Canada, along with its companion SDH or Synchronous Digital Hierarchy in Europe and most of the rest of the world.

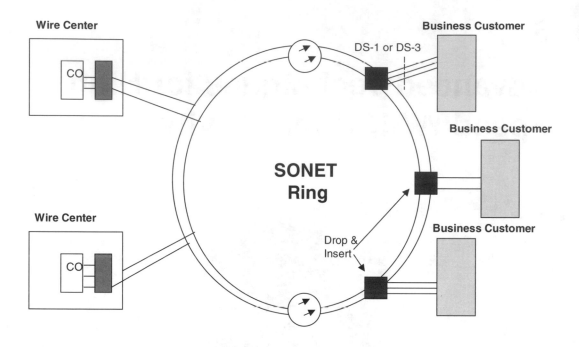

Figure 6.1: SONET access ring

Is SONET a good idea? The answer depends on your perspective. However, the question is much like a person posing the question "Is T1 a good idea?" in 1980. Considering the fact that T1 has been the backbone transmission medium for the United States and Canada for the last 40 years, then in hindsight the 1980 question: "Is T1 a good idea?" *earns a resounding YES!*

This is basically because transport technologies have a long life and do not become obsolete quickly. Still skeptical? Forget technology for a moment and consider the following example. Let's "fast-forward" to the year 2020.

Ask yourself this question, 20 years from now, will we still have railroad trains?

The answer is probably yes. In fact, most people will probably agree, there will be trains 20 years from now.

Now ask yourself: What will they run on? Will they run on diesel? Will they run on methanol? Will they run on nuclear fusion? Now the picture becomes cloudy. Therefore, although we can be sure that the basic technology will still be in use 20 years from now, we are unsure of the specifics. How about trucks? Will we have trucks 20 years from now? Once again, most people will answer, yes. What will they run on? What will their range be? Will electric vehicles proliferate? Will there be a breakthrough in battery technology? Once again, the details become hazy. Most people will agree that 20 years from now, we will still have trucks. So what does all this mean in the world of technology?

In information companies, telecommunications services are the equivalent of railroads and highways. And although the details are equally as fuzzy in the world of technology as they are in the world of trains and trucks, we can make a few basic assumptions.

In the world of railroads, for example, although we are not sure what *types* of trains will be running 20 years from now, *we can be reasonably sure that they will run over the same tracks*. And although we are not sure what types of trucks will be running 20 years from now, we can be reasonably sure that the interstate highway system will still be intact and that vehicles will still utilize it. Therefore, in the world of telecommunications, we can rationally hypothesize that fiber optics will still be the dominant means of transporting information at the basic level and that SONET will still be the backbone transmission medium. After all, we got a 40 year ride out of T1. We should expect at least 20 out of SONET.

In fact, this appears to be true. No one is envisioning a replacement for SONET at this point. By the time SONET is replaced, many of us will be wandering around shopping malls, looking for the perfect frozen yogurt, and asking each other why the kids won't call. In other words, SONET *affords us a 20 year planning window*—and how often do we get something like that in our business? This means it is probably rational to begin planning your backbone network now and that the network should encompass SONET technology. Okay then, given this perspective, where does ATM fit in?

ATM or Asynchronous Transfer Mode promises to be as revolutionary in the world of telecommunications as fiber optics was 15 years ago. ATM is the electronic equivalent of the trains on the tracks. Let me reiterate. The railroad tracks represent SONET technology. ATM represents the trains.

If you own the railroad, you only run trains when there is a product to be delivered, right? Similarly when you own a SONET carrier, you will only run ATM cells when there is information to be delivered.

How often can a railroad run a train? In the case of a railroad, the information carrying capacity of the tracks is far more than what they utilize. If trains run every hour on the hour, one can assume a given number of physical products, measured in tons, could be delivered. What if trains ran every *half* hour? Then conceivably twice the number of physical products could be delivered over the same tracks. What if trains ran every fifteen minutes? Then, of course, four times the physical product could be delivered. What if they ran every ten minutes? Every five minutes? Every single minute? *What if a train could run every few milliseconds?*

This becomes a ridiculous example when talking about railroads. However, ATM mapped over a SONET carrier can dispatch the electronic equivalent of a railroad train every few milliseconds. Are you starting to understand the correlation here? In this regard, the information carrying content of the underlining tracks, that is, the SONET carrier, becomes enormous, since all users technically share the bandwidth and run ATM "trains" only when information needs to be moved. This is beginning to happen already today

AT&T, for example, uses an ATM backbone for its Frame Relay network. What is Frame Relay? Briefly speaking, it is a modified fast packet switching technology. Frame Relay is the hottest telecommunications technology available today. But what is it really? Once again, let's draw back to our example of railroads and trains. In that context, Frame Relay would represent an 18-wheel truck.

Can someone drive a truck across country to deliver products? Sure, it happens all the time. However, we also see a lot of trucks riding flatbed cars on trains. This is because economically, it is more justifiable to transport this material by train rather than paying for diesel fuel and drivers. It is all a question of economies of scale.

Imagine a truck that drives to a railroad depot in Washington, D.C. It drives up onto the flatbed car of a train, is transported to Los Angeles, then drives off and finishes the trip to its destination under its own power. Something very similar happens in the Frame Relay world.

Many customers utilize Frame Relay in Washington, D.C. They install access lines to the AT&T POP (Point of Presence). From there, although it is not visible to the customer, AT&T actually puts this traffic on a transcontinental ATM link to Los Angeles. In Los Angeles, AT&T then strips the Frame Relay frames back out of the ATM cells and delivers the information in Frame Relay format to the user in L.A. All the user ever sees, however, is the truck.

Since AT&T has the economies of scale afforded by many Frame Relay customers, ATM quickly becomes cost justifiable as the medium of choice, especially for service providers, just as trains quickly become cost justifiable as the medium of choice when moving physical products.

So let's summarize.

- SONET is represented by the railroad tracks which are there all the time and are a large fixed physical investment. The carrying capacity of the tracks is measured by how many trains move over them in a given interval of time.

- ATM is the trains, and the more trains that run, the greater the number of physical products that can be carried down the line.

- Frame Relay is the electronic equivalent of a truck driving onto the ATM backbone when justified, due to cost or volume.

So what is the bottom line? We do not know what trucks will look like 20 years from now. We are not entirely sure what trains will run on 20 years from now, but we can be reasonably sure they will fit on the same railroad tracks. *With this in mind, it is never too early to begin building the tracks.*

Many companies today are taking exactly this approach. The advent of telecommunications reform allows us, for the first time, to put high volume fiber optic SONET service directly into customer locations. This is sometimes known as "reverse co-location." This is a prudent move since companies can be reasonably sure that the SONET backbone will be useful to them far into the future. What about ATM? ATM is still a relatively new technology and the operative question today, whenever anyone says ATM is, "Whose ATM?" When we start talking ATM and even Frame Relay, things get a little hazy. So the most rational course of action right now, until standards solidify, is to concentrate on what we can be pretty sure will be around for a while, mainly, the backbone SONET network.

Section 6.2 will concern itself with how a "Fortune" class user can install huge pipes for data, voice, and multimedia traffic, and how it can derive the competitive and strategic advantages such a network will afford. It will also describe a new regulatory and procurement process for securing these services and building strategic advantages that are difficult to duplicate by competitors and will increase a company's competitive posture in the long term. It will also deal with other opportunities afforded by telecommunications reform, including, the SOHO (small office/home office) environment and the convergence of entertainment (cable) and telephony companies, and will offer concrete solutions and examples of how to exploit each.

To the extent possible, we will attempt to use analogies to better illustrate the points to lay persons as well as technologists, since innovators in the field of advanced telecommunications may not be conversant in the acronyms and subtleties of the technology itself.

Some of the ideas presented will deal principally with cost reduction in the immediate term, since opportunities attendant with competition will abound. Others, perhaps most, will concentrate on the strategic advantages afforded by a 21st century telecommunications infrastructure. For now, consider some of the following points as a precursor to the details that will be presented later. Some are things you have already heard of, others represent the opportunities of tomorrow. How many of these concepts have you heard of before? *How many of the following is your company looking at right now?*

6.2 "Fortune" Company Checklist for Success

- Is your company considering any custom tariff filings such as AT&T Tariff 12s?

- Is your company shifting focus to LRIC (Long Run Incremental Cost) methodology in its negotiations with carriers? On the big things, OC3 and higher, start negotiations based on their costs, not an arbitrary number filed in a tariff. Tariff numbers contain a lot of other things besides the carrier's cost.

- Is your organization developing the capability to provide higher capacity services which support "native LAN" connectivity at speeds of 10, 16, and 100 Mbps?

- Are you providing or preparing to provide higher capacity for advanced applications such as televaulting, channel connection, video, and multimedia?

- Are you acquiring the capability to "burst" during periods of heavy network traffic, without affecting network performance?

- Does your company develop "shrink-wrapped packages" of enabling software for advanced telecommunications services, which can be easily assimilated, packaged, and sold by your sales forces, adding value to bandwidth?

- Have you trained your company's legal and contracts departments on the subtleties of negotiating custom telecommunications contracts?

- Has your company developed a telecommunications *regulatory affairs* capability to file custom tariffs on your company's behalf?

- Does your firm monitor and positively influence new legislation and regulatory rule-makings, to give your company the advantage of knowing about enabling regulation first?

- Have you tapped the potential market in *local* telecommunications services on your internal customers' behalf? (In other words, have you taken advantage of paradigm shifts in the telecom marketplace, principally, local competition and "telecom reform" legislation?)

- Who is the "800-pound gorilla" in your organization? Does your firm know how to present newly acquired services and capabilities to influential customers? Do they understand the new tools and capabilities they have available, and how to integrate them with the core business?

- The concept of a "Tariff 12" for the local telephone company (often representing 50 percent of network costs) is an exciting and marketable concept. It will be an effective "hook" in bringing back many of your "more autonomous" users! Are you considering it?

- Have you ever considered outsourcing telecommunications, not internally, but *outside* your company, capitalizing on the advantages you have created?

- Do you know the advantages of getting your customers "addicted to gigacells?" In other words, have you made transferring data as easy as using the phone? Since advanced networks tie in directly to the company's profitability (just like long distance calls produce tangible results like increased sales), the approach virtually assures a happy customer!

- Have you developed the capability to "meter" the telecommunications network just as many companies "meter" mainframe usage, allowing for equitable and accurate departmental charge-backs?

- Do you update your overall telecommunications infrastructure through targeted capital investments, aimed at specific core business goals? Can you measure the results in traditional ROI (return on investment) terms management understands?

- Have you ever considered establishment of a nationwide OC12 (622 Mbps) SONET network with drop-and-insert nodes at major points of concentration? (This must be pursued using LRIC methodology to avoid being cost-prohibitive.)

- Have you, or would you like to update your company's present T3 access facilities to SONET-based OC12, based on LRIC pricing? *Now?*

- Would you like to buy the best network management system available, including diagnostic hooks into the advanced network described above?

- Would you like to overlay advanced disaster recovery capabilities as an integral component of the new network, to avoid additional expense later?

- Would you care to sharpen your internal consulting capabilities and make fundamental structural changes to the telecommunications environment and its relationship to the core business groups, such as sales, production, operations, and others?

- Do you support an "emerging technologies" liaison to the core business units?

- Are you interested in looking for the "Microsoft Windows for Telecommunications?"

This last one is a long shot to be sure, but how often does a company even get that? The methodology described later in this text with give your company a better-than-average chance of realizing such a goal, while producing innumerable benefits in the meantime. Face it, the next revolution will be in advanced telecommunications. Nobody is the Microsoft Windows for Telecommunications yet; it is still out there for the taking by a foresighted and enlightened company willing to stretch the envelope. *It could be you.*

The "breakthrough" can be in any one of dozens of business lines. Will it be

- Banking and televaulting?

- Video and multimedia services?

- Distance medicine?

- Entertainment?

- Outsourcing?

- Manufacturing?

What does your company do, and what would it do if access to almost limitless bandwidth suddenly became a nonissue? Consider Figure 6.2 as you craft your "Seamless Solution!"

You are bound only by your creativity and imagination in moving toward the goal of being a true bellwether company, with technological capabilities unmatched by anyone.

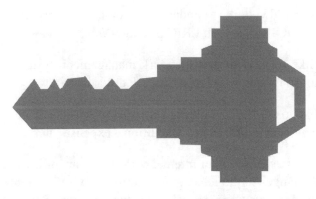

Mainframe Connectivity   Higher Speed Bandwidth   Video!

Detect / Diagnose / Fix / Control   PC Implementations

Higher Device to Personnel Ratios   Televaulting

Quality   Proactive vs. Reactive   Systems Integration

"Native LAN" Connections   Take Over Entire Processes

SO WHY NOT USE THE NETWORK AS THE INTEGRATING FACTOR TO UNITE USERS INTO AN ENTERPRISE-WIDE SOLUTION!?

Figure 6.2: Connectivity, standards, and network management are the keys . . .

6.3   SONET Technology

It is going to be SONET. As we already mentioned in the previous chapter, this technology, designed to run over fiber optic services, will be the backbone transmission medium for at least the next 20 years. After all, we got 40 years out of T1, right? How often do we get that long a planning window for technology implementation? Almost never! Besides, look at some of the other functionality SONET provides as an integral part of its design:

6.3.1   Fault Tolerant Ring Topology

Companies spend a lot of time designing fault tolerant networks and pay a hefty bounty for diversity to avoid single points of failure; and still cable cuts are an all-too-common occurrence. Much of this effort will become part of the network itself, as the network becomes smarter and "fixes itself" in cases of major failure, in much the same fashion that switched telephone service does today.

6.3.2   Allows 10, 16, and 100 Mbps Circuit Sizes

Once again, why do we use T1? Answer: Because that is all the telephone company has! This is changing dramatically. Now instead of us trying to shoehorn our data into what is convenient for the telco, they will begin custom tailoring for us! Virtually any size circuit can be mapped into a SONET frame. The same is also true for ATM. Many cost prohibitive applications like remote channel–attached mainframes, on-line vaulting, and other bandwidth hungry applications are well within reach, cost effective, and maintainable.

6.3.3   Drop and Insert Capability

Consider a transcontinental T3 (672 channels) circuit from New York to Los Angeles. Did you know that to drop off a single circuit in Denver you would need 57 separate boxes? That is because the T3 must be broken to T1, and the T1 to individual channels to get access to the channel you want. Then it must be multiplexed back up again into a T3 before continuing on to Los Angeles. SONET carriers have the capability to extract a single channel without breaking down the transport carrier. Needless to say, the cost benefit here is remarkable not only for carriers but users as well. The "Fortune" level customer in the U.S. today is already planning nationwide OC3 (Optical Carrier level 3 or 155 Mbps) rings with drop and insert nodes in major points of concentration.

6.3.4   Advanced Diagnostics

How many times has your network control center or help desk chased an intermittently impaired T1, when the problem only lasts a few seconds, then clears for a day, only to repeat the following day? SONET will literally be able to look down the line and ask, *"Okay, who was the wise guy who made the errors?"* And you know what? An intelligent repeater somewhere in Utah will answer back, *"I did it. Looks like you got me!"* The control center will then remove the offending component and restore perfect service. Contrast this to the half dozen technicians who become involved now every time a problem of this sort occurs.

6.3.5   Ability to Burst

Later, when users map ATM over SONET things really start to get interesting. For example, wouldn't it be great to have a red button that says, *"go faster,"* which you could press to temporarily turbocharge your network when the system begins to die under a severe load? When we begin mapping ATM over SONET transport systems, you will, and you may pay only when you do it. As users become "addicted to gigacells" in the conduct of their business, a pay-as-you-go pricing scheme will probably become the norm, just like in long distance.

6.4   How Do We Begin?

So how are we going to afford all this capacity? First, a company must become comfortable negotiating with carriers based on their long run incremental cost (LRIC) of providing the service, not purchasing through the standard tariffs. Next, these carriers must be induced to install network nodes on site with the user. The abused term for this arrangement is "reverse-co-location." In colloquial terms for the user, it means "make me a carrier too." The approach makes sense. If you were a factory, you would push for an entrance ramp to a major highway. If your product is information, your company obviously needs the electronic equivalent.

This new network-centric focus necessitates a radical transformation in the way you acquire network services, and how you will utilize them in enterprise networks of the future. It also means exploiting competitive opportunities in the changing regulatory scene, to garner maximum impact for your company.

6.5   Using Advanced Networks to Manage Client-Server Computing

Think about all the issues and questions that can be raised, namely:

- **Information Sharing—Insatiable User Demands for Capacity.** Users want to share larger and larger files, taking a greater and greater toll on network capacity.

- **Help Desk Support.** *What* help desk! We cannot ever attend enough schools to learn everything that is out there!

- **Workstation Support/Device to Technician Ratio.** Are there enough technicians in the *whole state* to support the number of devices I am planning to roll out next year?

- **Shared Capacity and CPU Cycles.** If I could find a way to broker the idle CPU cycles in my distributed environment, I could retire.

- **Controlling Software Installations.** Who cares? Let 'em buy what they want but don't call me when it crashes.

- **System Backups and Archival.** No problem. You can trust users to do backups. (I cannot even write this without laughing.)

- **Safety from Theft and Tampering.** The janitor hasn't spilled Clorox on my server yet. Besides, where else am I going to put it besides the closet?

- **Network Management.** "When it breaks I'll fix it—oh . . . it's a T3?"

Okay, enough funny stuff. The key points you need to derive from the above items are the issues that have haunted and taunted client-server deployments and until now, have not been adequately addressed:

- Capacity

- Help Desk

- Support Ratios

- Resource Sharing

- Software Controls

- Workstation Backups

- Physical Security

- Network Management and Troubleshooting

- Business Resumption Planning

These are not laughing matters. Everyone experiences these concerns to varying degrees. It just so happens, however, that a solution may be at hand, even for vexing problems like these.

So what will change this? *Connectivity, that's what!* And, on an unprecedented scale. With telecommunications competition will come new services, some of which will permanently alter the way we do business.

6.6   New Operating Modes Using SONET

When you really think about it, client-server is an inherently *inefficient* way of providing network support. However, it is the only way we presently have to put ergonomics, flexibility, and functionality on the desktop of the user. What happens, however, when telecommunications connectivity becomes available to *project* the same functionality, in a cost effective fashion? The answer is a return to a type of (dare I say it) *centralized* processing—and an increased level of network management capability.

It will not happen overnight, but it is silently changing now. Today, bellwether users are beginning to *partner* with service providers to design advanced, *customized* SONET networks, surgically tailored to the company's specific core business. Here is why.

6.7   Superior Network Management, Security, and Control

There are sound business reasons for implementing a "Johnson Space Center" level of network management and control. First, a superior network management system can be leveraged as a unifying factor to reclaim "renegade" LAN managers and provide a generally higher level of efficiency in the company by allowing users to concentrate on core business issues rather than technology. This includes providing (or selling) high level network management services to client-server customers who are essentially "black boxing themselves to death." Many information executives and end users alike are having second thoughts regarding management of technology, with sentiments such as:

- Can we really afford to chase technology?

- Do we really *want* to chase technology?

- Wouldn't it be better to focus on our core business instead?

Therefore, anything we can do to take the focus off maintenance, repair, and troubleshooting and allow the end user to concentrate on core business issues (the things that bring in money for the company—like selling things!) will be an exercise in enhancing efficiency and the bottom line. The message to the end user: *"We'll* manage the network, *you* concentrate on building Chryslers." It is a powerful pitch, if you can really pull it off.

Many companies already have configurations comparable to the one illustrated in Figure 6.3. This particular one was developed for a computer outsourcing company with a large base of remote users, and a desire for a cost effective way to outsource client-server.

In order to take full advantage of the options we have discussed, the company proposed a 622 Mbps SONET ring, perhaps with on-site ATM hubs after ATM standards solidify somewhat more. The basic transport fiber optic ring will be installed on a custom basis through a special project with one or more local service providers. Relax, it is SONET, and it is going to be here for 20 years. The ATM part becomes trickier with the operative

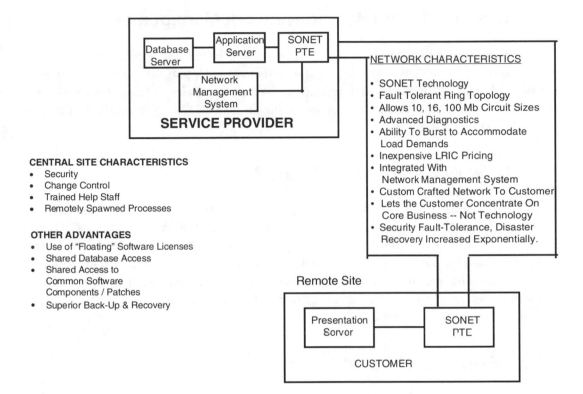

CENTRAL SITE CHARACTERISTICS
- Security
- Change Control
- Trained Help Staff
- Remotely Spawned Processes

OTHER ADVANTAGES
- Use of "Floating" Software Licenses
- Shared Database Access
- Shared Access to
  Common Software
  Components / Patches
- Superior Back-Up & Recovery

REMOTE SITE CHARACTERISTICS:
- Low Maintenance User Device
- High Device To Support Personnel Ratio
- More Concentration On Core Business, Less On Technology

Figure 6.3: Availability of advanced connectivity will change future network dynamics.

question, "Whose ATM?" The SONET part is a sure thing though. It is kind of like a user in 1980 saying "Is T1 a good idea?" It is hard to mess up with basic transport technologies. Besides, the newly competitive incumbent and competitive local exchange carriers (ILECs and CLECs) are dying to put SONET in for us!

6.8   Greater Reliability and Network Management

Since this network will fast become a revenue impacting system, we need to build in bullet-proof protective systems as well. This will probably involve tapping into network diagnostic data directly from the carrier, and interfacing a Johnson Space Center level of network control, as discussed earlier. Fault protection switching is also mandatory to protect against the all-too-common "backhoe fade."

6.9   How Does One Get SONET?

At the time of this printing, provisioning anything larger than a T3 generally requires an ICB or Individual Case Basis assembly. You generally cannot open a tariff book and order an OC12 in most areas. Additionally, the deal may also involve regulated carriers, so the Federal Communications Commission, state public utility commissions, or both will also get involved. That means beefing up regulatory support teams and extolling the virtues of this unique network in terms of company growth, jobs, technical prowess, and anything else except a means to save a buck or two.

6.10   Bring Your Wallet

Lastly, we will have to sell the boss on making a large capital investment by articulating the business vision to myriad departments, both internal and external, filing paperwork required at the state and federal level to register as a lobbyist, and fulfilling other concerns.

      Sound complicated? It is. But think of the potential payoff. If it is difficult and complicated to initiate, it is all the more difficult for your competitors to copy. That means your company enjoys the strategic advantage longer before others jump in. What is in it for the carrier? They stand to benefit most of all. The application you develop with them can be licensed, shrink-wrapped, and sold again to another customer (after you have moved on to something even better I should add.)

      That sure puts the carrier in a different class from those peddling long distance service at yard sales and swap meets! Incidentally, the more lucrative, tantalizing, and broadly adaptable the application is, the more accommodating you can expect the networking partner to be. This will be very important when you get down to the point of up-front concessions— who will buy what equipment, and what it will ultimately cost.

6.11   The Grand Strategy

By now you have realized that we are describing a radical new process demanding many new skills. Since whole books can be written on just a single one of these topics, we have selected eight specific strategies you can use to bring your custom network to reality. While they do not represent everything, they will get you well on your way! In order to keep improvements and capital expenditures consistent with your company's mission through effective linkage to its core business, I suggest establishing a detailed methodology for graceful integration of technology like the list shown below and in Figure 6.4.

1. Conduct detailed business process interviews.

2. Develop specific "feature packages" of services based on business need.

3. Divide packages into "existing" and "emerging" technologies.

4. Negotiate price on existing technologies in traditional fashion.

5. Develop "short list" of potential emerging technology partners.

6. Establish executive liaison and joint venture with selected partner(s).

7. Negotiate price based on LRIC.

8. Implement new technologies, share access to labs.

9. Release accomplishments to press. Both partnerships and emerging technology garner high investor interest.

6.12   Other Important Points

- Shift your company's focus to LRIC (long run incremental cost) methodology in negotiations with carriers. Start negotiations based on their wholesale costs, not an arbitrary number filed in a tariff. Tariff numbers contain a lot of other things besides the carrier's cost. Use a facility-based CLEC when possible. Their interconnection agreements will contain wholesale costs for unbundled network elements, forming the basis for negotiations based on real costs. Buy from CLECs using their resale margin on local services.

- Enhance and augment your company's shift into client-server computing through deployment of advanced network technology. Develop the capability to provide higher capacity services which support "native LAN" connectivity at speeds of 10, 16, and 100 Mbps. Use SONET as the underlying technology so you can change delivery systems later if required. Remember, they *all* will map over SONET.

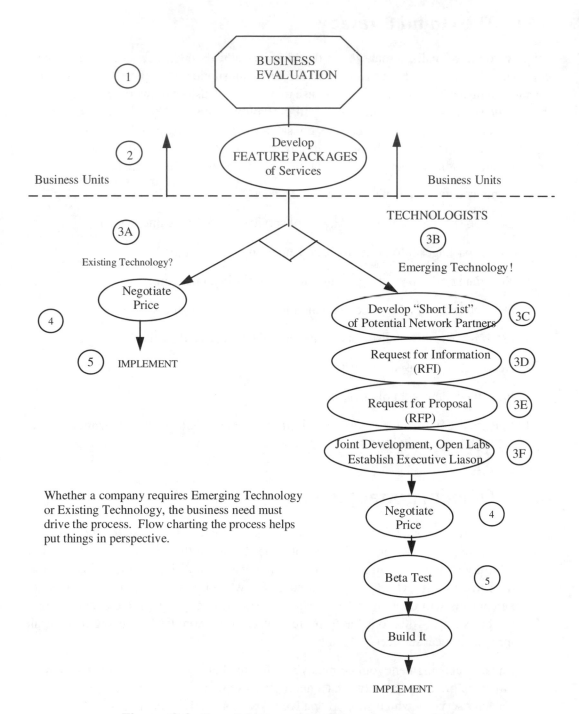

Figure 6.4: Graceful integration of emerging technology.

- Provide higher capacity for advanced applications such as televaulting, channel connection, video, and multimedia.

- Acquire the capability to "burst" during periods of heavy network traffic, without affecting network performance. That probably means ATM over SONET to provide that "expandable pipe" to deal with peak load demands.

- Develop "shrink-wrapped packages" of these advanced services, which can be easily assimilated, packaged, and sold by your partner's sales forces—or yours! Earn more money, as well as more prestige as the innovator!

- In addition to training your technical staff in negotiating based on LRIC pricing methodology, also train your company's legal and contracts departments on the subtleties of negotiating custom telecommunications contracts, given these new subtleties. Develop a telecommunications regulatory affairs capability within your company to file custom tariffs on your company's behalf. Monitor and positively influence new legislation and regulatory rule-makings. Give your company the advantage of knowing about enabling regulation first.

- Present newly acquired cost data and newly acquired services to your internal customers. Make sure they understand the new tools they have available.

- Update your overall telecommunications infrastructure through targeted capital investments, aimed at specific core business goals. Then measure them using an established ROI methodology. Organizations, like a telecom or IT department, that contribute, and prove their worth in an understandable format, are respected and rarely downsized.

6.13   The Promise of ATM

For those of you desiring a detailed technical overview of these technologies, you might consider the following section as a helpful refresher on ATM, and refer to the previous chapter on SONET. As stated earlier, ATM may be as revolutionary to telecommunications in relative terms as SONET was. Warning, these sections have a lot of bits and bytes. Policy and nontechnical individuals might want to consider a more basic overview or consult internal technologists on the details.

6.14   "Rewiring" Frame Relay—One Promise of ATM

Everyone has enjoyed the popular comedian Tim Taylor from the TV show *Home Improvement*. What is Tim Taylor's favorite saying in this show? He likes to look at something and

BYTE #	BIT 1	BIT 2	BIT 3	BIT 4	BIT 5	BIT 6	BIT 7	BIT 8
0	GENERIC FLOW CONTROL				VIRTUAL PATH IDENTIFIER			
1	VIRTUAL PATH IDENTIFIER				VIRTUAL CHANNEL IDENTIFIER			
2	VIRTUAL CHANNEL IDENTIFIER							
3	VIRTUAL CHANNEL IDENTIFIER				PAYLOAD		CLP*	
4	HEADER ERROR CONTROL							
5 - 52	PAYLOAD							

*CLP = CELL LOSS PRIORITY

Figure 6.5: ATM cell structure.

say, *"This would run much better if it only had more power."* And what does he do? Why he *rewires* it.

In a sort of lighthearted way, these sentiments hold true when users look at improved methods of interconnecting high speed LANs over a wide area network link. Users look at their network and make much the same statement, that things would run much better if they only had more power, and they too want to rewire them.

"Rewiring" telecommunications services is easier said than done. In the past, even the largest users have had to choose from a fixed menu of services, like garden variety T1s and T3s defined in telephone company tariffs. You could have one of these, but they are expensive and often underpowered for today's 100+ Mbps networks. Ask for anything other than these and you are basically out of luck. After all, the phone company is not Burger King. It is hard to "have it your way."

This is the part that is getting ready to change. With the advent of competition in local telecommunications, if a user cannot get what he or she wants from their incumbent telephone company, *someone* (like a new competitive upstart company) will probably be inclined to build it for them. This creates new opportunities to save money for certain, but more importantly, users can now "have it their way" in the deployment of new technology and services.

These unprecedented changes in telecommunications, fostered by the Telecommunications Reform Act of 1996 bode well toward the fact that our wide area network capabilities will match this impressive local environment.

For example, why do LANs and WANs have different rules? Actually, they might not for long. When the local incumbent telephone company or a competitive upstart telco installs a SONET terminal in your company's basement, you will bury T1 forever. When ATM cards drop in price below Fast Ethernet cards, you will retire Ethernet as well. Sounds ambitious? It is only a matter of time.

This section will provide some of the "whens and whys" of how your traditional network will begin to be replaced by new and exciting technologies, brought about in large measure by the dramatic changes in telecommunications regulation. (See Figure 6.5.)

These changes, as the figure illustrates, will manifest themselves in both the local and wide area networks, since ATM will be the first widely available technology that lives equally well in either world. Let's start with the LAN.

Imagine Tim Taylor describing his latest adventure in rewiring his local area network. "I started off with a measly little 10 base-T Ethernet, but I *rewired* it. Now I have a 622 Mbps ATM network over a SONET backbone carrier ... AURGH, AURGH, AURGH!!!"

So is 10 Mbps Ethernet outmoded? Is it obsolete? For most of us probably not. When one considers the business applications run by the preponderance of users out there, such as word processing, spreadsheets, and e-mail, 10 Mbps is quite adequate.

The problem is that you are not the only one using it. Naturally, the more people on a LAN segment, the more congested it becomes and the slower the overall performance becomes, as we discussed in previous chapters. How many times during the day have you heard someone yell out this phrase, "Why do they have to run this in the daytime?" What this usually means is one user becomes kind of a resource hog on the network, much to the chagrin of other users. For example, if an application involves primarily interactive transmissions over the LAN, such as in a customer service environment, 10 Mbps works fine, even for numerous users. The reason for this is, not all users transmit at the same time. Sometimes they are on the phone talking to customers, sometimes they are on break, or sometimes engaged in other tasks. In fact, in many environments a user may actually be on line only seven to ten percent of the time, even in a busy customer service environment. This means many users are sharing the same bandwidth, but at different times. When all users yell out in the familiar chorus "Why do they have to run this in the daytime?" it is usually because one user starts making large file transfers or consumes an inordinate amount of the resources available on the LAN at that time. This is where the problem begins. In the 10 Mbps Ethernet environment we cannot create more capacity, but we can segment what is there. That has been the modus operandi for about the last 10 years. A whole team of "filter kings" constantly monitors LAN performance, usually staying just one step ahead of the wrath of huge bands of yapping users. But just suppose users had access to virtually unlimited bandwidth on a virtually instantaneous basis. This is the promise held out by ATM. ATM will make data communications like the phone. Moreover, ATM will offer true isochronous capability—suitable not only for data but for voice and video as well. Let me explain.

Imagine you are doing a full motion video transfer of me, the author, at one of my speaking engagements. Typically a video link is isochronous, meaning that the bits leave the location while I am speaking and arrive at the location where you are viewing in sequence with each other and with the timing relationship maintained. Another example of isochronous would be a voice telephone call. Once again, in order to move an intelligible voice from one end to the other, the words that leave the location where I am speaking must arrive at the location where you are listening in the same order. Imagine in the case of the video conference and in the case of the voice call, if the video and the voice were broken up and all of these packets arrived at the far end out of sequence with each other. Imagine further that other users could jump in the middle and send their data at the same time, thereby delaying packets in my transmission from one end to the other. This would mean the voice coming through on the other end would be unintelligible since the words would be out of sequence, phrases would be mixed, and everything would be out of sync. The video would be even funnier, as you watched me go from one end of the room instantly to the other end of the room, up on the ceiling, down on the floor, and every which way. I'm being a little lighthearted with this, but you get the idea why isochronous capability is important. LANs are nonisochronous since they are shared by a number of users. For data, momentary transmission delays are no problem. To the user, they appear as slightly lengthened response time. For isochronous capability however, such as voice or video, most

LANs will not work for these transmissions.

ATM is a high speed cell switching technology which promises to revolutionize both local and wide area telecommunications. In fact many experts believe that the impact of ATM will be as revolutionary in scope as the advent of fiber optic communications were on telecommunications in general.

ATM is designed primarily for speed—T3 (44.736 Mbps) and up. But applications are being written for ATM as low as T1 (1.544 Mbps). How does this compare with that other popular service, *Frame Relay*?

Well for starters, 75 percent of Frame Relay deployments today are at 56 Kbps. Frame Relay itself is designed to operate up to and including about T1 speed. However, new applications are being written for Frame Relay that will take it up to about T3 speeds. The only real overlap between the two, therefore, is between T1 and T3 speeds, and even here it gets hazy.

Frame Relay is a nonisochronous packet switch technology, meaning it is not ideally suited to voice or video. Many users today are running voice over Frame Relay to be sure. However, it is the type of proposition where you are taking your chances. If you do not mind apologizing for the quality of your voice, go ahead and run it over Frame Relay. Frame Relay is ideally suited for data, however, since data traffic is delay tolerant and does not require isochronous transmission. Frame Relay is also one of the hottest selling, most widely available, and most cost efficient technologies available today. ATM on the other hand is more limited in availability. The promise of ATM however is the ability to provide an isochronous form of transmission suitable for voice, data, and video, with the same pricing dynamics and efficiency of Frame Relay. How will it do this? Check out Figure 6.6, then let's walk through an example.

Let's assume we want to take this Frame Relay network as shown in Figure 6.6 and make it suitable for isochronous traffic. Remember that in transmitting any data from A–C for example, each packet of data may take a different path through the network. Therefore the data packets from A, which are numbered 1-2-3-4-5, may in fact arrive at location C in the order 5-3-1-2-4. This does not present a problem for data, since higher level protocols like TCP/IP will rearrange the data making sure there are no missing packets, and deliver the data to the premises equipment on the other end with an acceptable amount of delay for a data user. But what if this is a voice call or video? If a video conference is encoded at location A as 1-2-3-4-5 and arrives at location C as 5-3-1-2-4, you can imagine what the video will look like. On the other hand, packet switching is an enormously efficient way of moving data around, since any facility with capability available the instant you want to send data can be utilized. This provides great facility utilization and actually allows the carrier's cost to serve the same number of users on a much lower number of circuits. So, if we want to make this suitable for isochronous capability, here is what we need to do. First let's start with the size of the links between the switching nodes and the network.

With the advent of fiber optic capabilities we can actually make these things far larger

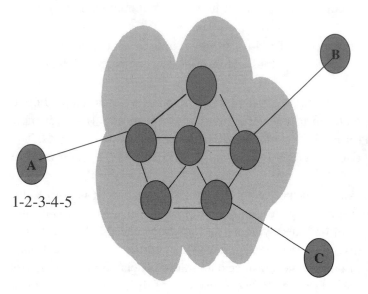

Video over Frame Relay poses problems. The non-isochronous nature of the medium can cause bits to arrive out of sequence.

1-2-3-4-5

5-3-1-2-4

What can we do about it? How can we "rewire" frame relay?

1. **Make links larger.**
2. **Make switches faster.**
3. **Make packets shorter.**

ATM does all of these to provide true isochronous capability!

Figure 6.6: Rewiring Frame Relay.

than they have been in the past. So the first thing we are going to do to this network is to make the internodal links far *larger*.

Next, we need to do something with the switches. Conventional switches are far too slow in propagating data. Therefore we need to develop a whole new switching technology including switches which are literally faster than hellfire and damnation. In order to make switches fast, we need to make them simple. So simple in fact, that they are more hardware than software and can literally be wired into the network. This is what we have done today with ATM switches. The fastest of these today are in excess of 10 Gbps or the equivalent of 45 T3s. That is fast enough to avoid significant delays in the network (also called latency), and again, speed the network up to the point where it could possibly support isochronous capabilities. So the next step is to make the switches *faster*. But we are still not done yet.

To make a network fast, truly fast, the size of the data packets needs to be small. We cannot have processors in the network waiting around for a large data packet to propagate without increasing latency. Therefore, we need to make the packets small and also consistent in size. If all the packets are the same size, and all the packets are small, they can be switched through the network quickly. So the third thing we need to do to turn a fast packet network like Frame Relay into an isochronous network like ATM is to shorten the size of the packet. In this case into a 53 byte *cell*. By doing these three things—making the links larger, making the switches faster, and making the packets smaller—we can indeed derive the advantages of isochronous transmission with the cost efficiency of a cell switched technology. This is the promise of ATM.

So getting back to the question posed at the beginning of the book, why do LANs and WANs have different rules? See Figure 6.7. Presently, a signal from a network interface card in a computer in New York to a network interface card in a computer in Los Angeles undergoes numerous protocol formats and translations along the way. Why? When ATM cards drop to the price of Ethernet cards, and when ATM services become widely available, the signal will never change from the network interface card in New York to the network interface card in California! LANs and WANs will have exactly the same rules.

It gets even better. If the two sites in New York and California are connected over a high speed SONET link (that terminal the local telco put in your basement, remember?) of say, 622 Mbps, a single user theoretically has the whole pipe to himself for a large file transfer. This is true bandwidth on demand. A 17 second ATM "call" can move a 4 gigabyte engineering diagram across the country. Now that is more power. Tim Taylor would be proud. "AURGH AURGH AURGH."

Therefore as users, it behooves us to consider the revolutionary changes occurring in the telecommunications environment as not only cost saving opportunities (even though they are), but also perfect opportunities to expand the capabilities of your networks in ways which have been heretofore unavailable. ATM is one area to consider to be sure. Have you any other good ideas? If so, now is the time to pose them to your service providers.

So who is the winner? ATM? Frame Relay? Probably the best way to make that de

TODAY
LANS and WANS have different rules

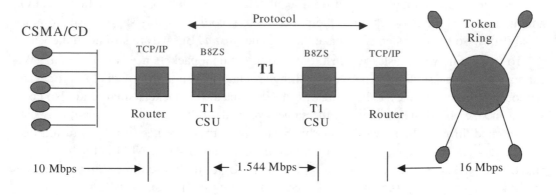

TOMORROW
LANS and WANS have the SAME rules

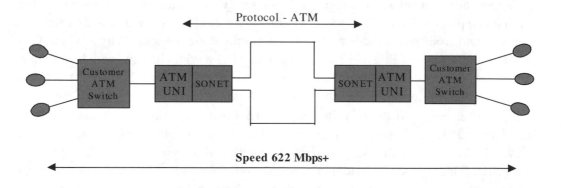

Figure 6.7: LANs? WANs?

termination is (once again) to consider the *business need*. If your organization moves a lot of multimedia or heavy-duty engineering diagrams (T3 speeds and above) or needs truly isochronous capability, ATM is probably the ticket. If your organization moves lower speed data only, the economics of Frame Relay become rather irresistable (see Figure 6.8). A review of isochronous characteristics from Chapter 2 is shown in Figure 6.9.

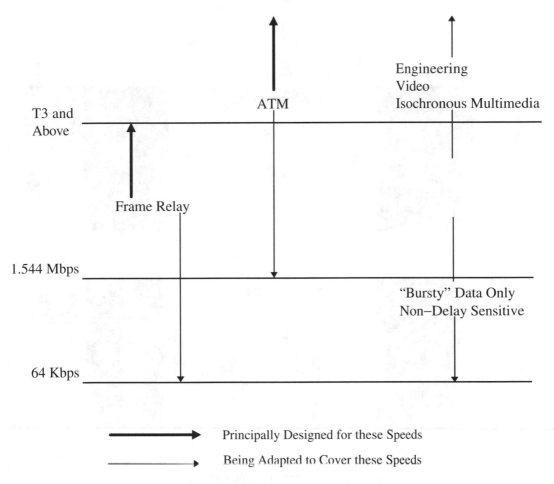

Figure 6.8: ATM versus Frame Relay—who wins?

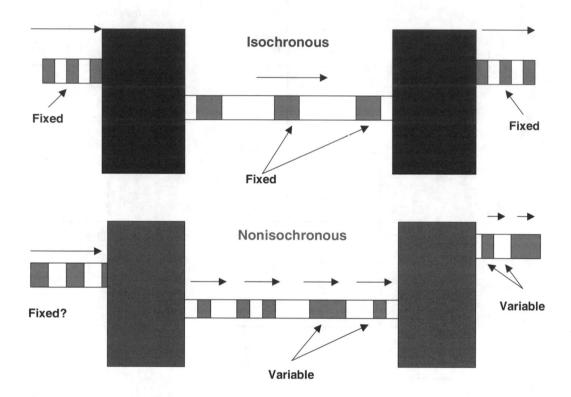

Figure 6.9: Isochronous/Nonisochronous.

7

Supporting Small Offices, Nomadic Users, and Telecommuters

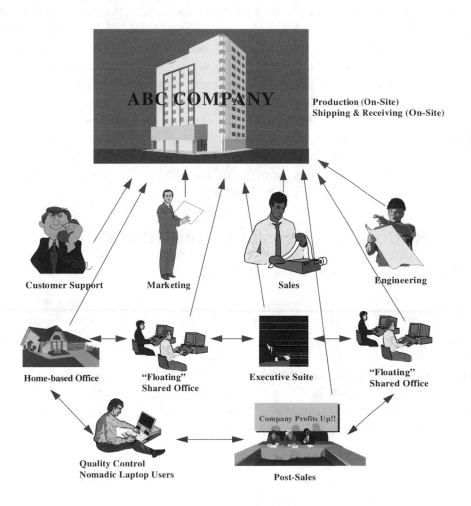

Figure 7.1: Close collaboration no longer requires close proximity

7.1   Introduction

What kind of user makes the "ideal" telecommuter? Realistically speaking, virtually any kind of job can be transplanted to a remote location as shown in Figure 7.1. The tough part is providing the tools at a reasonable cost. First and foremost, an understanding of what the user *does* is a prime consideration before meaningful implementation of any program can take place. Start by splitting people into two categories:

1. Production Workers

2. Knowledge Workers

As Figure 7.2 shows, either kind of worker can work remotely. Each, however, will have differing technology requirements. Production workers, as the name implies, are dead in the water without enabling technology, but on the other hand, are easy to standardize and support. Knowledge workers, on the other hand, need very specific package requirements. These workers can usually do something else however, even if the technology fails them. Standardization is important for help desk support as seen in Figure 7.3. I realize this is a very broad-brush overview, but this kind of understanding will have major implications on your telecommuting program, and your ability to support it in the future.

The word "telecommunications" means something akin to "communicating from afar." It seems funny therefore that when people who are nomadic workers, or telecommuters, approach the corporate telecom department, they are told, "We can supply you with telecommunications services, but you have to be *here*!"

7.2   Cost Considerations in Virtual Offices

"We Are Going To Save 'Tons of Money' "

If there was ever a reason *not* to telecommute, this is probably one of them. Although savings are certainly probable if approached properly, if that is the motivation for an employer, the project is probably doomed from the start. Savings are usually recognized over the *long* run in ways such as reduced employee turnover and increased productivity.

True, "upfront savings" are enjoyed by some employers. For instance, an expanding company may be in a situation where it *must* move to larger quarters, or send employees off site simply due to lack of room. The expenses associated with moving, as well as the increased rent for a larger space, may very well be much higher than setting up employees to work from home.

Still, an employer who is not willing to invest to properly set up an employee for telecommuting—someone who is going to cut corners on equipment, services, or supplies—is likely to create an environment where it is difficult for the employee to be productive from home or any other remote location. Setting up the proper workspace for a

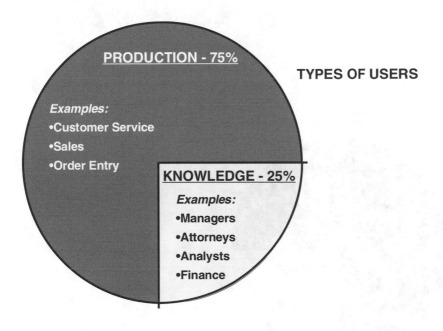

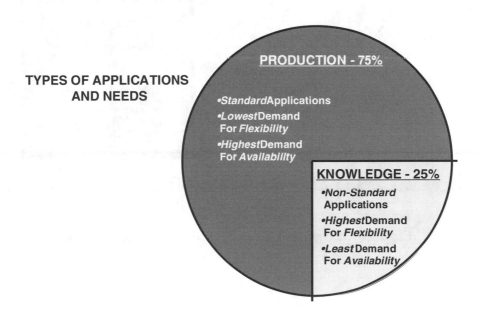

Figure 7.2: Managing nomadic users (understanding user requirements).

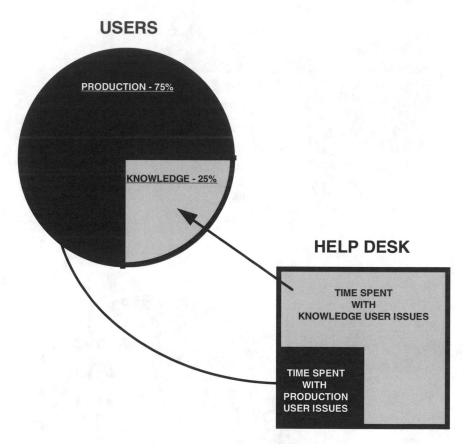

BOTTOM LINE:
STANDARDIZATION IS IMPORTANT!

Figure 7.3: Help desk support versus type of applications.

telecommuter or nomadic worker sometimes requires significant up-front costs. Expenses must often be encountered by either the employee or the employer for fax machines, copy machines, personal computers, phone lines, and more. Even somewhat mundane items such as office supplies can add up. The physical space must be available and have proper lighting. It should be set up with air conditioning and heating, and be away from the noise and commotion of the house. Even seemingly small improvements such as multiline telephones can be problematic in some locations, such as apartments and condos that may be wired for only one line. All of these considerations will be discussed in greater detail later in this text.

The real cost savings associated with a telecommuting environment include (Figure 7.4):

- drawing a higher quality employee

- increasing that employee's productivity

- reducing the employee's absenteeism

- retaining quality employees (turnover is one of the biggest expenses for many employers)

- lower overhead costs

- accommodating employees with physical disabilities

Many of today's employees, especially professional employees who are measured more by the quality of their production, work well under this system. Sure there will be a few that will abuse the system. However, a cooperative arrangement which gives employees greater control of their time will generally reduce their stress, greatly enhance their home life, and cause their productivity to skyrocket—all because you treated them like adults and simply trusted them to do their jobs.

7.3 Getting Employees the Tools They Need at Home

For successful support of nomadic users, this is the name of the game. It is very easy to take items like telephones for granted, but consider people who live in condominiums or apartment complexes. Many of these dwellings may be wired for only one or two phone lines and the landlord may have some very serious misgivings about making changes. The telecommuter may also want to set up an office in an outlying area with inferior telephone facilities that will not support data calls or in an area where even procuring a second telephone line is difficult or impossible due to a shortage or lack of facilities.

Services that are common in the metropolitan areas may be next to impossible to get in the outlying areas. For example, there are still central telephone offices in the U.S. today

Advantages
- **Increase Productivity**
- **Reduce Stress**
- **Support our Families**
- **Lower Overhead Costs**
- **Reduce Absenteeism**
- **Draw and Retain Higher Quality Employees**
- **Help Environmental Protection & Energy Conservation**
- **Comply with the Americans with Disabilities Act**

Disadvantages
- **"Always On Call"—Invasion on Private Life**
- **Isolation from Business Culture**
- **"No Water Cooler"—Lack of Social Contact with Co-workers**
- **Lack of Nonverbal (Face-to-Face) Communication**
- **"No Help"—Lack of Support Systems from the Work Place**
- **The Workaholic & Burnout**
- **Distractions**

Figure 7.4: The virtual office.

that do not offer touch tone telephone service. There are even some areas where getting something besides a "party line" is next to impossible! Fortunately, these sorts of problems are being rapidly resolved.

Still, services within serving areas can vary widely. What about the telecommuter who would like a distinctive ring so that he or she can differentiate between home and business calls? While this is a very common feature in the metropolitan areas, it is not available in many outlying areas. A preliminary call to the telephone company inquiring about available services would also be advisable before buying expensive equipment that depends on more modern technology.

7.4 What About Communications?

In addition to telecommunications, routine office communications can be complicated somewhat by having a large nomadic work force (see Figure 7.5). What about routine memorandums or industry magazines that circulate within the office? (How about the Far Side Calendar page that is circulated in the department daily?) Some types of interdepartmental communications that are paper based (and perhaps too large to be easily faxed) present a challenge when trying to keep a large mobile workforce well informed. This is a minor concern to be sure, but one more thing that must be addressed to support the work at home environment.

7.5 Which Jobs are Best Suited for Nomadic Operations or Home Officing?

Editor/reporter, programmer, medical transcriptionist, administrative writing service, customer service representative, sales representative, telemarketing representative, engineer/designer, architect, financial analyst, claims representative, real estate broker, technical writer, graphical artist/typesetter—all of these positions can potentially be handled by a nomadic work force, but that is not to say other positions are not equally suited for home officing (see Figure 7.6).

There are a number of issues to consider when evaluating a position as suitable for home officing. Will the worker need access to files or records that are only available at the office? Is it cost effective to duplicate those records for the home office? Could the records be computerized or made available to the telecommuter in some other fashion? Does the position require face-to-face interaction in the office? Are there frequently last minute instructions that need to be conveyed face-to-face? Could the need for office files and face-to-face interaction be handled if the nomadic worker spent a specified time each week or month in the office?

♦ What about Zoning
  Laws?

♦ What about
  Telecommunications?

♦ What about
  Communications?

♦ Will Managers of
  Telecommuters (or
  Telecommuting
  Managers) Need to be
  Retrained?

♦ What about Tax
  Implications?

Figure 7.5: The virtual office, other considerations.

Frequently, jobs most suited for telecommuting are those jobs that evolve around information management. That is to say, the best candidates are those jobs that require sending, receiving, manipulating, analyzing, and/or presenting information. "Information" is becoming more and more of a hot commodity in our world. Positions that are potentially suited for telecommuting are also generally jobs where the computer and telephone are indispensable work tools. With technological advances, the ability to easily handle information from afar is becoming easier and easier as well. Consequently, we are seeing an increase in workers who work at some place other than the traditional office. You can count on exponential increases in the numbers of telecommuters in coming years.

- ➢ **Editor/Reporter**
- ➢ **Programmer**
- ➢ **Medical Transcriptionist**
- ➢ **Administrative Writing Service**
- ➢ **Graphical Artist/Typesetter**

- ➢ **Customer Service Representative**
- ➢ **Sales Representative**
- ➢ **Telemarketing Representative**
- ➢ **Engineer/Designer**
- ➢ **Architect**
- ➢ **Technical Writer**
- ➢ **Financial Analyst**
- ➢ **Claims Representative**
- ➢ **Real Estate Broker**

Figure 7.6: Jobs most easily adaptable to the home.

7.6   Enabling Technologies for Telecommuting and Remote Access

We have talked a lot about the who, what, when, where, and why of telecommuting. Now let's talk about the "how." The short answer is *communications*. Of course, that is a little simplistic or we probably would not even be writing this book!

Communications today has dramatically changed from communications of even a decade ago. Yes, the phone systems have changed a lot. Still, the biggest change is probably the way we communicate and share data. Many use various types of technology to share data as well as voice daily. We are turning to technology to help us more and more in our day-to-day work activities.

Given our increasing dependency on technology in our lives, time spent learning about the options available today (as well as those expected to be available in the near future) can be time well invested. We encourage anyone who is in the process of developing a corporate telecommuting policy to slow down long enough to thoroughly evaluate options available through present and emerging technologies.

7.6.1   Dial-Up Data

How fast is fast enough? For most users, the answer is "just a little bit faster than what I have right now!" It is surprising to think that as recently as 1978, state of the art for data communications was 1,200 bps, roughly 1/20 of today's rather ho-hum 28.8 Kbps modems. Even so, anyone who has surfed the Internet at 28.8 Kbps (and watched graphical information drag along) knows that faster is better. The problem is that, generally speaking, our public switched telephone network dates back to the days when Sarah, the operator, was asking "What number can I connect you to?"

Although I may be exaggerating somewhat with this comment, it is almost true. Telephone service has changed little in the last 100 years. The primary purpose of the public switched telephone network is for verbal communication. Data is a secondary consideration and requires a lot of outboard solutions developed over the last 20 years or so. One is the modem.

7.6.2   Modem Technology

The word "modem" stands for "modulator/demodulator." It is essentially designed to take a digital signal from a personal computer, for example, and change it into an analog wave-form, which will transverse a network designed to talk over. State of the art at the time of this writing is about 33.4 kilobits per second, although some vendors have modems capable of speeds up to 56 kilobits per second (asymmetrically).

However, modems are becoming old news. Even at 33.4 Kbps, they are beginning to reach the limits of what a standard telephone channel can provide. Additionally, a lot of components in the network (which we must remember was designed for verbal communication and not data transmission) adversely affect the transfer of data. For example, a telephone company will often put loading coils in a voice circuit to eliminate envelope delay distortion and other impairments on the line—great for voice, but bad for your modem. Such action means the souped-up, high-priced modem you just bought will not work at its full potential simply because of constraints on the public network.

Such limitations can be significant since telecommuters work just about anywhere. For example, have you ever received a call from the 802 area code? I did the other day and could not imagine where it would be from. It is not a new area code. It has been around for a while. The call happened to be from a publisher working on a mountaintop in Vermont. Vermont is a beautiful state and there are lots of reasons people emigrate from the hustle and bustle of New York and Boston to work on mountaintops in Vermont. Environments like these, while beautiful for quality of life, can be a complete backwater with regards to telecommunications. Can you imagine what types of communication facilities go to mountaintops in Vermont? Such conditions provide significant challenges to publishers who depend on high speed data connections and Internet access. You can work with the telephone companies to upgrade and improve service. We have done that before. However, it is an adventure! I wonder what would happen if this publisher ordered an ISDN line to his mountaintop retreat?!?

7.6.3 Modem Manufacturers

Manufacturers of modems are almost too numerous to mention. The big names, of course, include Motorola, U.S. Robotics, Packard Bell, IBM, AT&T, 3Com, Cardinal Technologies, and many others. The average cost of a high performance modem these days is around $100 and dropping. Contrast this to 12 years ago when a 9,600 bps modem only worked on a four wire private line and cost around $8,000!

7.6.4 Modem Standards and Speed

With terms like V.22, V.32, V.34, and so on, the standards for serial communications used by the majority of modems are not an easy thing for the average user to understand. All of these standards for modems are set by the ITU (International Telecommunications Union) which decides the speeds and protocols for standard devices that communicate over the telephone lines. These standards also include other items such as error control and compression.

The most common speeds for modem communications are associated "V-dot" standards and are as follows:

V.21	300 bps
V.22	1,200 bps
V.22 bis	2,400 bps
V.32	9,600 bps
V.32 bis	14,400 bps
V.34	28,800 bps

Standards also cover error correction and comprehension which effectively increase throughput of a modem. The most widely known of these are the V.42 and V.42 bis data compression standards. The V.42 protocol can be used with any modem from 1,200 bps and up to compress data, thereby increasing the throughput of the modem. For example, under some conditions, a 2,400 bps modem can yield transfer rates of up to 9,600 bps when uncompressed data is transmitted. Similarly, a 28.8 Kbps modem can yield up to 115.6 Kbps more, roughly a ratio of 4 to 1. However, if the data that you are transmitting is already compressed (such as in a binary file or a document compressed with PKzip or some other zip software), the compression ratio is far lower. For example, document files compress very nicely and yield close to the 4 to 1 compression rate.

In round figures, a 2,400 bps modem using a V.42 protocol can be expected to send the equivalent of about 450 characters per second. On the other end of the scale, a 28.8 Kbps modem transmitting or receiving an uncompressed file could do so at the equivalent of approximately 3,000 to 3,300 characters per second (cps).

For a very rough estimate of throughput, take the speed of a modem and divide by 10, provided compression is not being used. In this fashion, the speed in cps for a 2,400 bps modem would be approximately 240. The speed in cps for a 9,600 bps modem would be approximately 960. This is simply because each character is 8 bits which, when combined with start and stop characters and other overhead, coincidentally works out to about 10. Remember, the estimation of modem speed reached by dividing by 10 does not take into account any compression, and is not a scientific calculation by any means.

7.6.5   Modem Configurations and Related Settings

Once you have selected your modem and are on line there are other things to consider. First, if you are using any type of high performance modem be sure that your communication port or serial port is set for the appropriate baud rate. Set it for at least 38.4 Kbps and up to 115 Kbps, if possible.

Second, check to be sure that the COM port you are using has a 16550A, 16550AF, or 16550AFN UART chip. The easiest way to do this is to exit Windows, go to the DOS prompt, and then type the letters "MSD" while in your DOS directory. This runs the Microsoft diagnostic utility which checks the configuration of your personal computer. When the main screen appears, select the "C" option that examines your COM ports. The type of UART chip used will appear at the bottom of the COM port's screen. Many computers use a 16450 UART chip which, while satisfactory for older modems, tends to fall down on the job when higher speed communication is used. You will not get the full mileage out of your souped-up 28.8 Kbps modem unless you get a COM port equipped with a 16550-based UART. Without getting exceedingly technical on the subject, the 16550 chips possess a FIFO (first in, first out) buffer capability, which really speeds things along on high-speed communications. You naturally have to make sure that your COM ports and interrupts are correct—for example, IRQ, IRQ4, and port 03F8 (for COM1, and so forth).

Next, you may want to consider upgrading to Windows 95, 98, or NT which would mean not having to hassle with a lot of this. Windows 95 also offers vastly superior COM port drivers than Windows 3.1 or 3.11. As an option to Windows 95 for simplifying setting your COM port configurations, you might consider an outboard program which installs over windows such as, Turbo Com. (This second option is preferred by old "DOS bigots" like me!!)

Finally, consult the manual for your modem to be sure you have set the right hardware flow control option and the appropriate AT commands for compression.

7.6.6   Other Issues Affecting Modem Communications

Hard disk drivers may affect modem communications by interfering with COM port drivers in the system. They can slow down your modem communications. For this reason, try and keep your hard disk drivers up-to-date with the last known "good" version for your particular hard drive.

7.6.7   What About the Phone Line?

The phone lines, in and of themselves, can be a real pain when it comes to modem communications. While some technologies today allow speeds of up to 56 kilobits per second on a standard dial-up line (with special conditions), some lines in outlining areas barely support 2,400 bps. After all, we are sending data over a network that is over 100 years old—and it was designed primarily for talking, not data transmission.

Numerous other things can affect data transmission as well. For example, loading coils on cable (designed to improve voice transmission) causes intermodulation distortion and greatly affects data delivery. Other types of "pair gain" systems also introduce intermodulation and harmonics as in some of the older versions of subscriber loop carriers used by the telcos. Such systems can literally cause fits for people operating at over 1,200 bps. Although much of this equipment is being phased out, much also still remains.

Our telecommuter on a mountaintop in Vermont, for example, may well be on an SLC40 type pair gain device relegating him to 1,200 bps per second. Sometimes this restriction can be relatively easily overcome.

In one instance, we added "line conditioning" to a residential line for a telecommuter working at home. The telephone company charged a one-time fee of around $60 and about

$1.50 per month thereafter. The work performed involved removing the data line from an SLC40 pair gain device (which was causing significant distortion) and assigning the circuit instead to a solid copper cable all the way back to the telephone central office. Since this was a distance of approximately 30,000 feet (standard telephone lines generally operate at 18,000 feet or less) a gain device was added to insure proper signal levels.

As an additional plus, the circuit was then assigned a special circuit ID number identifying it as a data circuit (presumably to insure special handling in case of trouble). This provided a good solid data communication solution, and was really a bargain when weighed against the cost of $1.50 per month. You may get a strange look at first when inquiring about "line conditioning" from the telephone company, especially if you reside in an outlying area. Be persistent! Once they figure out the engineering hurdles, the improvement will be worth the trouble.

7.6.8 Consider an Internal Modem

Modems, particularly high performance internal modems and ISDN terminal adapters, often exhibit superior performance in the internal versions. This higher performance is achieved because many of the dynamics of converting information to serial synchronous data that can pass through the COM port are bypassed. An internal modem provides an additional benefit by freeing up a COM port that could be used for other purposes.

7.6.9 Don't Overlook the Obvious!

One thing often forgotten by modem users is to temporarily disable the call-waiting feature of their telephone service. With older modems, the call waiting tone causes the connection to drop completely. With newer modems that have more advanced error correction features, the tone still slows the connection down significantly.

By simply adding "*70" to your modem's dialing string, callers dialing your modem number will receive a busy signal instead of your connection receiving a disruptive call waiting tone. If you have pulse dialing rather than tone, add "1170" in the dialing string. After you terminate your modem connection, your call waiting service will resume as normal.

7.6.10 Providing Internet Access

The Internet is becoming an indispensable tool to telecommuters and roving nomadic users. The ability to tap into a worldwide network for e-mail, file transfers, and information gathering on a casual homebound or nomadic basis is almost unprecedented and is revolutionizing the way home officing is viewed. A typical dial-up Internet account costs anywhere from $15 to $20 per month on either a flat or usage sensitive basis. ISDN connections are more expensive, averaging $30 to $40 a month on a flat or usage sensitive basis.

Dedicated connections to the Internet are also available in many areas and may not be as cost prohibitive to a homebound telecommuter as one might first think. These usually command a one-time installation fee of $400 to $600 and a monthly fee ranging from $200 to $300. These prices are high, but cost tends to pale somewhat to the intensive user who really needs this type of capacity, such as an engineer.

With so much in print these days about the Internet, there is little need to go into an extensive discussion on what the Internet is. It has revolutionized virtually every office place throughout the U.S. and worldwide. As such, homebound and nomadic users will demand similar access as Internet-based applications move into the nontraditional office place as well. These demands will be filled in the immediate term by dial-up connections to the Internet at 33.4 kilobits (or less) per second. In the near term, the need will be filled by more emerging technologies such as ISDN and cable modems (addressed in this chapter). It would not be a stretch to say that Internet access is the single most decisive factor behind the demand for the new high bandwidth services.

Can you imagine the torment of trying to surf the Internet or paint today's graphical images at 300 bps? The surprising thing is that this was state of the art until about 1968. What a difference 30 years makes! But we do not have to look back that far to see that we have indeed come a long way.

As late as the early 1980s, 9,600 bps was pretty heavy-weight stuff, available only on dedicated, specially conditioned private line circuits. Data speeds that could be reliably switched over the public network with dial-up modems were much lower and limited to 2,400 bps as late as 1985 or so. Many 2,400 bps modems are still in use today, but those who use them, especially for large file transfers, realize that this speed is completely inadequate.

By the late 1980s, higher speed switched modems came into existence. Even so, today's state-of-the-art devices exceeding 28.8 Kbps over a regular telephone circuit, drag along when they have to push graphical images.

Even the modem itself is now an endangered species. This is because the trouble with modems is that they run over the same old hog—the 100+ year old analog technology of a network designed to carry voice telephone calls. Everything in the future will be digital, and it has already been happening.

7.7   The Future is Digital

In the 1980s digital technologies began to proliferate. Users were first offered speeds of 56 or 64 Kbps, first on a dedicated basis, and later on a switched basis. Even though these higher speeds are available today, their availability is still often limited. More recently, however, ISDN (Integrated Services Digital Network—a 15+ year old standard) finally began coming of age. Since it only requires the same old copper wire that is already out there, but carries a digital signal, ISDN will be the vehicle (for the telcos at least) that brings digital technology to the masses. We will discuss ISDN in detail in upcoming chapters, but here is a quick explanation for now

ISDN consists of two 64 Kbps circuits plus one 16 Kbps signaling circuit. As you can see, having an aggregate of about 128 Kbps is preferable to being limited to 2.4 Kbps for people who transfer large files or move lots of graphics. Internet surfing is both a pleasure and an adventure at these speeds. Of course if that is not enough, you can always still order a T1.

7.7.1 T1 Services

When users began screaming for additional bandwidth, the telephone companies looked around at what was available within their network, and that just happened to be T1. The only reason we use T1 today is essentially, *it is the only thing the telephone companies had!*

A T1 is nothing more than a 1.544 Mbps data circuit, which was primarily designed and intended to carry voice telephone calls. During the 1960s, T1 came into widespread use within the public network when the telephone company was looking for a cost effective alternative to putting a separate set of wires for each circuit between each of their switching centers. Therefore, the T1 standard was adopted to allow 24 individual talk circuits to ride on the same set of wires between the telephone company's switching centers. T1 was not originally intended as an end user service, it was strictly designed to be confined to the telephone company's network. T1 was also not originally designed for data service. Rather, it was designed to carry multiple voice channels intended to be talked over at the same time.

A classical T1 multiplexer breaks the 1.544 Mbps T1 into 24 individual circuits. Each of these circuits in turn can run analog data at speeds up to about 28.8 Kbps, and digital data at speeds up to 64 Kbps. Today, everyone and their dog uses T1. It is not, when you think about it, a terribly convenient speed. Our networks run at 10, 16, 100 Mbps, everything but 1.544 Mbps. We were forced to take what we could get though, and that was the standard T1 rate. Today it is still not enough. The next step after T1 is a giant step to T3—quite cost prohibitive, especially for the homebound user!

7.8 Users Will Drive the Future

Many services that are beginning to become available, such as SONET, Frame Relay, SMDS, and ATM, take the opposite approach to the one we discussed earlier. Rather than being primarily designed to carry voice and then adapted for data, tomorrow's networks are being designed to carry data with secondary provisions to carry voice. We are no longer limited to what the phone companies happen to have. Today, the users, not the telcos, are driving the process, and dozens of new services are being developed and refined for specific user needs.

These new technologies will, in large part, be driven by the needs of leading edge end users of telecommunications services. *And not a moment too soon!* Users are ready for the change.

7.9   ISDN: Innovations Subscribers ~~Don't~~ *Do* Need!

ISDN has been a frequently maligned service, which is finally coming of age. The notion of "Innovations Subscribers Don't Need" (also interpreted as "I smell dollars now" by others) has been replaced by "Innovations Subscribers *Do* Need!" in just a few short years. We will cover what has changed, and how we can avail ourselves of a host of new advantages presented by this first generation of affordable digital capacity to homes and offices.

ISDN, which incidentally really means *Integrated Services Digital Network*, is a 15-year-old standard for carrying two 64 Kbps "bearer" channels, which can be used for data or voice, along with a single 16 Kbps "delta" channel for signaling and other data. That is why you often hear of it referred to as 2B+D. Since ISDN only requires the same old copper wire that is already out there, but carries a digital signal, ISDN will be the vehicle (for the telcos at least) that brings digital technology to the masses.

The Internet has been the principal "killer application" that has finally brought ISDN to life. In fact, in many parts of the country today the local telcos are doing a land office business in this technology. And why not? Having an aggregate of about 128 Kbps is at least five times faster than the swiftest modems on the market today.

In addition, a user can also talk over one B channel and send data at 64 Kbps over the other. Using this approach, a user can point to items on a screen using a program package similar to Intel's Proshare, while conversing with another user at the same time. This is possible with modem technology as well, but it can be clumsy and slow at modem speeds. ISDN really speeds things up. Additionally, anyone using H.320 compliant video equipment can send moving images at speeds as low as 64 Kbps. At that speed they are not really high quality, but are passable, especially when reduced to a small matchbook size window on the user's screen. At 128 Kbps, with today's technology, it is not network broadcast quality, but does not look bad at all for desktop conferencing applications. Needless to say, traditional Internet and BBS (bulletin board system) file transfers over ISDN are a breeze in comparison to slower modem transfers.

ISDN has come along not a moment too soon. It is almost comical to contemplate the fact that for years, if you made an interexchange telephone call, you were essentially speaking over a switched 64 or 56 Kbps data circuit! The technology has been around forever, the problem is, the applications for it were not. Only comparatively recently has the network been modified and packaged for users to transmit data. A 128 Kbps ISDN call uses the equivalent of two 64 Kbps trunks. That is why the telcos were able to adopt it with relative ease. It really is not all that new technology-wise. Already, however, there are alternatives on the horizon that may shorten ISDN's heyday.

Cable companies are attacking the Internet access market with a vengeance. Many of these devices provide speeds of 10 Mbps or more. We will discuss some of these technologies later in this book. For the immediate term, ISDN is the easiest way to bump up to digital speeds, albeit modest speeds by tomorrow's standards, with relative ease.

As we stated earlier, ISDN service is delivered to the home or small office on two wires, but the connection is full duplex, meaning data can be sent each way simultaneously. The B channels carry either telephone conversations or data at speeds of 56 to 128 Kbps. The D channels carry signaling information, such as call set up, call progress, and calling party identification information. D channel ISDN has also been proposed for such items as Automated Teller Machines. They provide a constant connection to the ATM for data, which is required. (The bank wants to know presumably if someone walks off with one!) ISDN is also good for multitenant environments which have only one set of wires, since two numbers are assigned to two B channels, all over only one pair of wires. Those of you who have ever lived in an apartment or condo know what a pain adding additional lines can be. The majority of BRI (Basic Rate Interface) ISDN applications are going into homes. Let's walk through a few of the steps required for a successful installation in a home or small office.

7.9.1 Check Availability in Your Area

ISDN is technically deployable today almost anywhere. The service is designed to operate within 18,000 feet of a central telephone office without a whole lot of headaches, over the same copper wire that provides your phone service. Not all telephone offices are, however, ISDN capable. Many telephone companies opt instead to make a relative few central offices ISDN capable, and remote-serve users in outlying areas.

For example, an ISDN circuit we had in an outlying office was actually served by a telco central office 25 miles away in downtown Dallas. (See Figure 7.7.) This is not a problem but it can cause you, the user, to pay extra charges for channel extension from a foreign central office or exchange. Indeed, one complaint by ISDN users who are remotely served is that they have to make a long distance call to the number next door, since their ISDN "dial tone" comes from a foreign switch miles away!

7.9.2 Select a Service Plan

A business telephone line generally costs around $30 per month. A switched 56 Kbps circuit costs up to $500 per month. An ISDN BRI line providing two 64 Kbps channels (and two telephone numbers) generally costs around $60 per month or $30 per line. A small price to pay for vastly superior capabilities. However, the price varies based on where you are. If the ISDN line is cheaper for a base rate, but carries a per-minute usage charge, any price advantage will be quickly used up. For example, think about the amount of time it is possible to spend surfing the Internet. Additionally, it will cost you to talk as well. Make sure you do an apples-to-apples comparison to avoid unpleasant shocks. Your local telephone company may offer you an option of flat rate or usage sensitive pricing, or it may mandate one or the other. In Texas, for example, ISDN from Bell is flat rated. ISDN from GTE, however, is usage sensitive. In California or New York you can count on it being different. Explore your options thoroughly before taking the plunge, and make sure you are on a service plan you can live with.

7.9.3 Select Required Features

One of the things that impressed and astounded me when I ordered my first ISDN line was the wide variety of features available. All of the standard telco features like touch-tone, caller ID, three-way calling, speed calling, and so forth, are available.

Additionally, other useful features are available as well. For example, the ISDN line will send calling party information to you over the D-channel. If you have a data call up on one channel, and are talking on the other, your equipment will display something like "Mother-in-Law trying to call in." Actually, it will display her number, and probably her name as well. This allows you to decide whether the caller is important enough to terminate one of the connections you have up and also effectively screen your calls.

A word of caution. The telco person that will be working on the translations (programming) for your ISDN line will be looking at four or five pages of options. Any one of these, if keyed in wrong, is often lethal. There were many times with our ISDN line when we called in a routine change like "make B channel 1 roll to B channel 2 when B channel 1 is busy" and we were down or impaired for days while the telco tried to make it work. Some of this is probably the natural learning curve by people who are still much more comfortable with telephone test sets than data. This is not intended to irritate the telco folks, but ISDN is definitely "not your father's Oldsmobile" and the learning curve is steep. The best advice would be to thoughtfully consider which features you will need *before* installing the line, and then try not to change them later unless truly necessary.

7.9.4 Select Your ISDN Equipment

Most units suitable for homebound environments or small offices have one or two analog ports which allow you to plug in phones, modems, and fax machines that you may already have in the office. This is an important feature. These devices also often support features such as Caller ID, call hold, transfer, and conference. ISDN units also boast sophisticated call routing capabilities and often allow you to direct different ISDN numbers to different analog devices such as phones, fax machines, or modems, almost like PBX DID lines. These are "uptown" features to a small operation and present a good image to the incoming caller or client. See Figure 7.7 for details.

ISDN "modems," terminal adapters, and other units geared more toward nomadic, roving, or homebound users are manufactured by the following companies:

3Com
Ascend
ADAK

ADTRAN
Alpha Telecom
Farallon
ISDN* TEK
Motorola
ZyXel

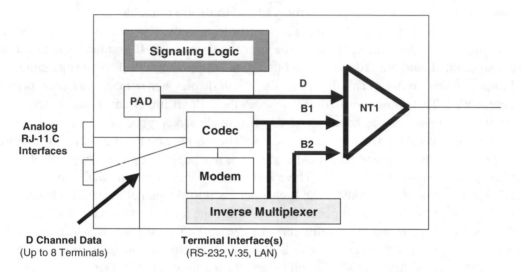

Figure 7.7: ISDN terminal adapter components.

If you expect to stay in a relatively low key environment such as a residential office, look for a device with an integral NT-1. It will save you hassle and money. If you want to connect multiple devices later and "split" the NT-1, you will probably want an outboard NT-1 which will give you access to this interface. Remember, in either case you will have to plan for power as described below.

7.9.5   Plan for Power Backup!

One commonly made ISDN error is forgetting to plan for power disruptions, unless you live in Europe where the telco takes care of the problem. In North America, ISDN power is the user's responsibility. That means investing in a low-priced UPS like you probably already have for your computer. Since ISDN terminal adapters draw very little power, you should be covered for a day or so on even a small UPS. Absent this precaution, install a standard phone line along with the ISDN. Since it is powered by the telco, it will work when the ISDN line will not. At least it is relatively inexpensive. The cheapest UPS designed for a personal computer (about 60 bucks) will probably provide eight hours or more of backup to an ISDN terminal adapter, since the power they require is so minimal.

7.9.6 *Place Your Order/Monitor Progress*

This part can be an adventure. ISDN is almost invariably provisioned by a special group within the telco. It is really amusing to hear a telco operator ask "Eye-Ess-Dee-Whut?" when calling Information for the ISDN provisioning number. (I am sorry this sounds so negative on the telcos, but it is true, try it! It is not my intention to be offensive, just to prepare you, the reader, for what will probably happen. If it goes well for you, congratulations— your local telco is ahead of the pack!)

Once you have the number, the telco will need to know the kind of equipment you have and a lot of details. Equipment manufacturers have really improved in providing these specifications in usable formats, and telcos have come a long way on their standards. (There, I said something nice about them.) If you order a garden variety configuration and do not specify options that are too weird, it should turn up and operate just fine. Ours did not work first time out, but came up smartly the following morning. The channel extension piece was not wired down in an intermediate office, but that was quickly resolved.

7.9.7 *Select an ISDN Capable ISP (Internet Service Provider)*

You will want to have something to connect to! Use the same criteria for selecting an ISDN ISP as for a normal one—capacity, customer service, price, and so forth. Remember, your hot new 128 Kbps connection will perform like a 2,400 bps modem if your ISP experiences frequent congestion problems. In fact, many people are rather underwhelmed by the performance since the Internet is inherently slow. If the server on the far end will only dish out 900 characters per second (about 9,600 bps) that is what you get, even though the link to the ISP is capable of more than 6,000 characters per second. When everything is working right, however, things really rumble. At 3:00 a.m. you feel like you own the net!

7.10   ISDN and Remote Access Equipment Examples

The following figures illustrate three different strategies for supporting nomadic users and telecommuters.

1. The first strategy (Figure 7.8) is "low dollar—easy entrance," such as might be used by a low-end residential user. It uses an inexpensive device called an ISDN terminal adapter, in this case the Motorola BitSurfer Pro.

2. The second strategy (Figure 7.9) is more complicated, versatile, and expensive. The "Jetstream" box illustrated is typical of some higher-end products for the one person office environment. It features integral voice mail, and "follow-me" capabilities quite useful to traveling salespeople for example.

3. Third, companies like Siemens and Nortel are introducing ISDN key systems (see Figure 7.10). These fit quite conveniently into the class of remote users that are too large for individual lines, but too small for T1.

These examples are for illustrative purposes only and will probably become dated quickly. Nonetheless, they do inspire ideas when considering equipment capabilities and making your selection.

> **GOOD VALUE (INEXPENSIVE)**
> **DYNAMIC BANDWIDTH ALLOCATION**
> **ANALOG PORTS**
> **CALLER ID, CALL FEATURES**
> **STANDARD "AT" MODEM COMMAND SET**

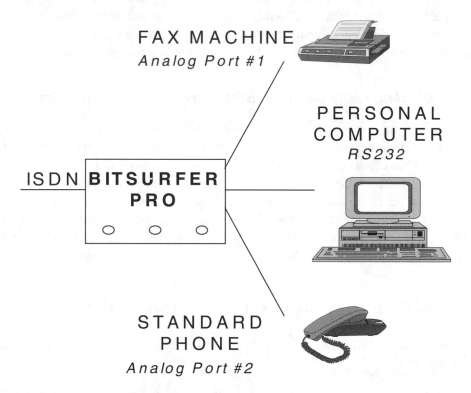

Figure 7.8: Motorola's BitSurfer Pro.

7.11   ISDN to the Internet

ISDN to the Internet is nothing new. What is new however, is ISDN access in the context of the recent changes in telecommunications reform. Could an ISP become a telephone

HOT IDEA!

SUPPORTS "TELECOMMUTER EXTRAORDINAIRE"

- ➢ Up to 16 Telephone Numbers

- ➢ Sophisticated Call Handling Capabilities

- ➢ Integral Voice Mail

- ➢ Fax Handling (Inbound, Outbound, Storage, and Remote Retrieval)

- ➢ "Find Me" Capabilities

- ➢ High Speed Data Communications

Figure 7.9: Jetstream Communications' The Front Desk.

company? In today's environment, you bet it can. We expect that by the time this book is in production, many ISPs will take this step, either by themselves or with the aid of a certified carrier.

Moreover, new technologies such as xDSL are displayed most efficiently by carriers. This means that close carrier collaboration, or outright certification by ISPs will be the order of the day.

7.12   ISDN! How to Do It

ISDN offers a reliable 64 Kbps or 128 Kbps connection every time—unlike the so-called 56 Kbps modems, which often connect at less than 28.8 Kbps. ISDN also gives you two telephone numbers on two separate paths. One can be used for your residence line, the other can be used for a home business. Either one will support your Internet access! Figures 7.11 through 7.14 give examples of what ISDN can do for you.

➢ ISDN TELEPHONE SYSTEM

➢ GREAT FOR SMALL OFFICES

➢ UP TO 3 ISDN LINES (24 NUMBERS)

➢ HIGH SPEED DATA COMMUNICATIONS

➢ ADVANCED KEY TELEPHONE FEATURES

Best Price / Performance for Quick Setup of
Nomadic Office (6 People or So)

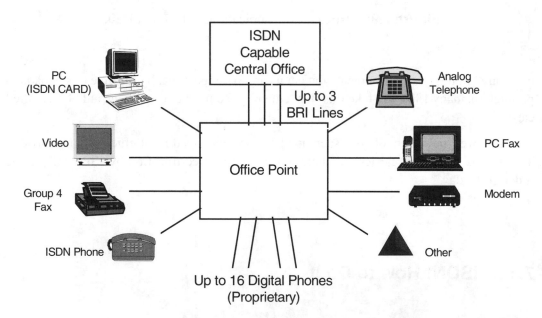

Figure 7.10: Siemens/Rolm's Officepoint Communication System.

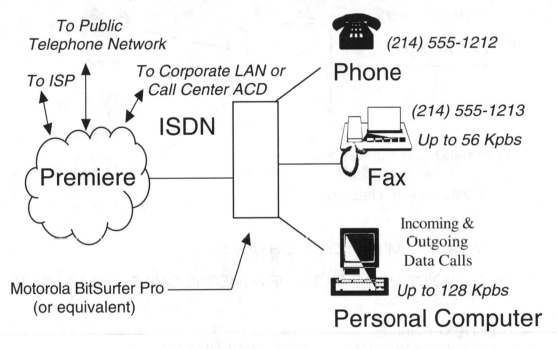

Any Two Devices Shown Below Can Operate At The Same Time

Figure 7.11: ISDN can be like three lines in one!

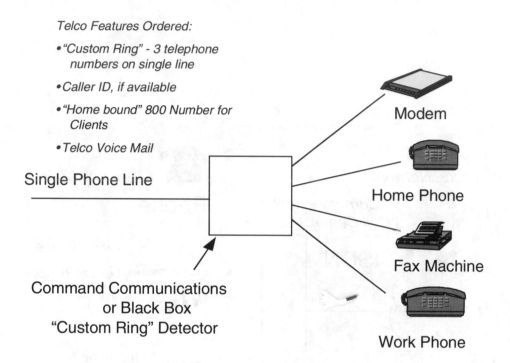

Telco Features Ordered:

• *"Custom Ring" - 3 telephone numbers on single line*

• *Caller ID, if available*

• *"Home bound" 800 Number for Clients*

• *Telco Voice Mail*

Single Phone Line

Command Communications
or Black Box
"Custom Ring" Detector

Modem

Home Phone

Fax Machine

Work Phone

➢ SEPARATE NUMBER FOR EACH DEVICE

➢ TELCO VOICE MAIL PICKS UP IF SOMEONE CALLS WHILE FAX OR MODEM IS IN USE

➢ IT'S INEXPENSIVE.
THE SWITCH BOX ONLY COSTS AROUND $100

➢ DISADVANTAGE:
FRUSTRATING TRYING TO SHARE ONE LINE.
CANNOT USE FAX AND PHONE AT THE SAME TIME.

Figure 7.12: ISDN example # 1.

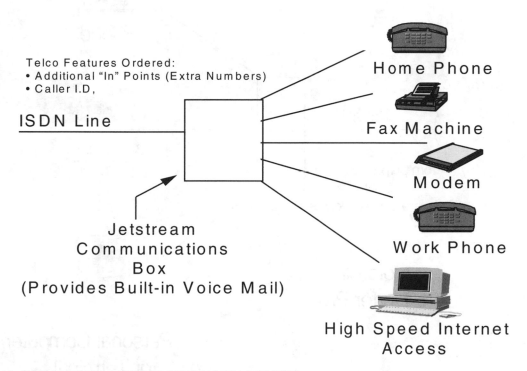

Telco Features Ordered:
• Additional "In" Points (Extra Numbers)
• Caller I.D,

ISDN Line

Home Phone

Fax Machine

Modem

Work Phone

Jetstream
Communications
Box
(Provides Built-in Voice Mail)

High Speed Internet
Access

➢ SEPARATE NUMBER FOR EACH DEVICE

➢ VOICE MAIL AND POWER PROTECTION BUILT INTO THE UNIT

➢ TWO ISDN PATHS ALLOW SIMULTANEOUS USE OF INTERNET AND VOICE

➢ "BIG COMPANY" IMAGE

Figure 7.13: ISDN example # 2.

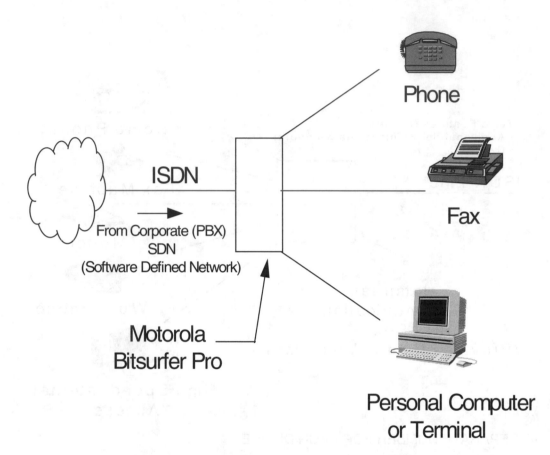

> ➤ CALLS DIRECTED TO HOMEBOUND USER BASED ON AVAILABILITY
> INFORMATION SENT TO NET VIA ISDN D-CHANNEL. CALLS ROUTED
> AUTOMATICALLY TO "NEXT AVAILABLE USER"

> ➤ HIGH-SPEED NETWORK ACCESS TO COMPANY ORDER SYSTEM

> ➤ FAX LINE PROVIDES OUTGOING CALLING WHEN REQUIRED, WITHOUT
> TYING UP INCOMING LINE

Figure 7.14: ISDN example #3.

7.13   Summary

That is about all there is to it. The rest is basically taking the plunge and learning the intricacies of ISDN provisioning through actual experience. While the goal of the service providers is to make ISDN as easy to order as a typical telephone line, collectively, we users and the telcos are still a ways from making that happen. Even so, the service is less complicated to provision than it was even a year ago. For example, many equipment providers provide turnkey sales, meaning that *they* spend their time on the phone if something goes awry, *you* do not. That should be an important consideration before purchasing equipment.

To recap, ISDN stands for *Integrated Services Digital Network*. ISDN is a 15-year-old technical standard designed to carry two 64 Kbps "bearer" channels, which can be used for data or voice, along with a single 16 Kbps "delta" channel for signaling and other data. That is why you often hear of it referred to as 2B+D.

Since ISDN only requires the same old copper wire that is already out there, but carries a digital signal, ISDN will be the vehicle (for the telcos at least) that brings digital technology to the masses. It also stands to reason that for this reason, for the immediate term anyway, ISDN will be the telephone company's medium of choice since they are obviously awash in copper facilities. But will the telephone company be the only show going with regard to delivery of high bandwidth services? Not on your life!

8

Emerging Competitive Alternatives for Small Offices, Nomadic Users, and Telecommuters

8.1 The Telecommunications Reform Act of 1996 and its Implications on the SOHO Market

New services, under new pricing, bundles, and terms—this was a large part of the intent of the Telecommunications Act of 1996. Thus far the act has been somewhat of a nonevent, but the pace is quickening and new services are beginning to proliferate, even in the small office environment. ISPs (Internet Service Providers), for example, are beginning to file for local carrier certification. At the low end of the spectrum, an ISP can expect that if it bundles e-mail addresses with the customer's telephone number, that it will reduce the ISP's "churn" rate (people changing to new ISPs). They are right. At the other end of the spectrum, carrier certification gives ISPs the chance to pioneer new technologies, like xDSL, by virtue of being able to buy the parts wholesale from the incumbent carrier and build them to specs. And this only considers ISPs.

The 1996 Telecommunications Reform Act will be remembered as the major catalyst for affecting the most far-reaching changes in telecommunications, at least since the AT&T breakup in 1984. Volumes can be written about this legislation alone. For purposes of this book, these are the most germane portions of what the legislation will change.

1. Telcos will be allowed to raid each other's territory. For example, it will be possible to get GTE service in Southwestern Bell territory, or Ameritech service in Bell South territory, and so on.

2. Long distance carriers will be allowed to raid the local telco's territory. Already, people are signing up for AT&T and MCI local telephone service.

3. Cable companies will be able to enter the telephone business, and likewise, telcos will be allowed to provide cable. Already, AT&T and others are marketing "Direct

TV" mini-dishes, and cable companies are marketing ultra-high bandwidth Internet connections and data transmission services.

These items only scratch the surface of everything that is in the new legislation, however, they stir interest among technologists like us, particularly in the areas of new, cheaper, and faster services which competition will bring. It will be quite a ride. Ten years from now there may only be three phone companies. And it might be quite a surprise to see who the survivors are! Imagine if the big three were AT&T, Time Warner, and British Telephone, for example! Oh well, that is getting pretty far out. For the near term, we will enjoy new exciting services that will complement or even replace the ever-more-available ISDN.

8.2   First of the "Emerging Technologies"—ADSL

What better service for a copper-rich local telephone company than one that utilizes copper? One of the major advantages of ISDN is the fact that it can use the same old copper wire that delivers your local telephone service for high speed (128 Kbps) data. Need even more speed? Then consider ADSL; it uses copper too. ADSL stands for Asymmetric Digital Subscriber Line. ADSL gives telephone companies a shot at the high bandwidth market previously thought to be the exclusive domain of cable companies.

Since ADSL uses the same telephone facilities as traditional telephone service, it allows use of existing analog modems, fax machines, and telephones while maintaining a high speed data connection at the same time. ADSL does this by carrying three separate frequencies over the same line. The first set of frequencies carries your telephone service, while another carries a 16 to 640 Kbps data signal (depending on the product utilized) for upstream information from your home to where you are connecting, such as the Internet. The third signal is a high speed, downstream connection, which runs anywhere from 1.544 Mbps up to about 9 Mbps. Like ISDN, ADSL is a digital signal.

The reason ADSL makes sense for most users is that we typically download much more information from the Internet than we transmit to it. Using ADSL, a 64 Kbps channel is used to "request" a download, which can be made at speeds of 1.544 Mbps or higher. ADSL could in fact be the Web surfer's dream. (See Figure 8.1.)

Therefore, you can think of ADSL as a small pipe for delivering your requests (requests sent to the Internet asking for information) combined with an ultra-large pipe used for receiving the requested information (back from the Internet). It can be cheap too. We have heard of an offering in U.S. West territory, for example, where ADSL is priced at $60–$100 per month. And remember, that also includes dial tone on the same line.

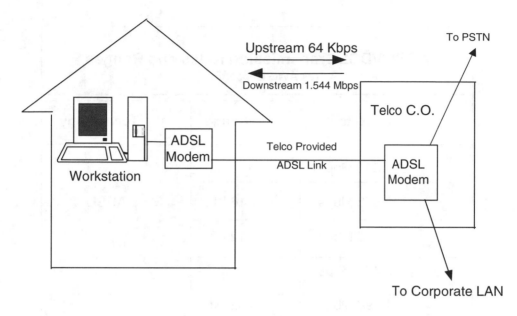

Figure 8.1: DSL to the home.

8.3   DSL Equipment

To use ADSL, once it is available, you will have to purchase an "ADSL modem." Just like with ISDN, this is a misnomer, since ADSL is a digital technology. (Please note, these are still very much under development.) A typical unit would have three connectors. The first, of course, would go to the telephone company interface. The second is a standard RJ11 telephone jack used to plug in your phone. The third is an RJ45 Ethernet connector, which connects the service to your personal computer or local area network. In other words, you can use your standard telephone service and a high speed Internet connection at the same time.

A number of field trials are underway at this time. For example, GTE in Dallas has launched one of the most aggressive trials of ADSL service, locally in public libraries, bookstores, and with some GTE employees. The equipment currently being tested includes products from Aware and Westell Technologies with transmit speeds from 64 to 500 Kbps and received speeds at 1.5 to 4 Mbps. Higher speeds are possible, however, with speeds of up to 9 Mbps available.

8.4   DSL Speeds and Distances

Speed and reliability of the xDSL signal is a function of cable gauge and length. Figure 8.2 shows the correlation based on today's technology. Unlike T1 which has repeaters every

ADSL/VDSL Transmission Rates and Ranges (24 AWG Copper Pair)		
Bit Rate	**Range**	**Technology**
1.544 Mbps	18 Kft.	ADSL
2.048 Mbps	16 Kft.	
6.312 Mbps	12 Kft.	
8.488 Mbps	9 Kft.	
12.96 Mbps	4.5 Kft.	VDSL
25.92 Mbps	3 Kft.	
51.84 Mbps	1 Kft.	

Figure 8.2: How fast is fast?

Traditional T-1

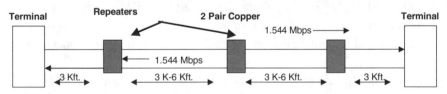

HDSL Link

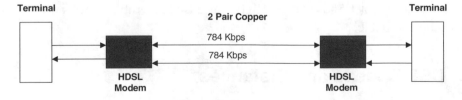

Figure 8.3: T1 versus HDSL configuration.

5–6,000 feet (Figure 8.3) and carries a digital signal, xDSL uses contiguous copper and carries an analog signal, much like today's modems. As such, it can be affected by many of the same impairments.

8.5 VDSL

For the truly bandwidth hungry, another similar technology referred to as VDSL (Very High Data Rate Subscriber Line) promises even higher speeds. VDSL can operate at rates of between 13 and 55 Mbps, but distance limitations are more of a factor than with ADSL. In the case of VDSL, a subscriber loop can only be approximately half the distance as with ADSL. Present technology limits the distance of ADSL to about 9,000 feet from a telephone company central office. This distance is greater than the maximum distance T1 lines allow (6,000 feet), but less than that for ISDN (18,000 feet). Even so, you can serve a lot of subscribers within this radius. Expect initial deployments of both VDSL and ADSL to be bundled with equipment packages by the telephone companies. That is another thing they are allowed to do now with the liberalization of the regulatory picture.

8.6 Cable in the SOHO Market

With the entrance of cable companies into the telecommunications business, one of the first offerings will be ultra-high-speed data connections available on a broad basis. This is not to say the market is there for the taking, or that things are rosy for the cable companies. Competition to these companies from every pizza-pan-sized satellite dish dealer is acute, and cuts right into the cable companies' core business. For this reason, they can be expected to embrace technologies that will encourage users to utilize their service, and high speed Internet access may be the "killer application" they are looking for.

8.7 ADSL as a Competitive Threat to Cable

ADSL is the most direct response the telcos have to a cost-effective network solution which rivals the bandwidth of coaxial cable. Indeed, one of the major advantages to this technology as we have already stated is the fact that it can use the same old copper wire which delivers local telephone service for high speed data. ADSL gives telephone companies a shot at the high bandwidth market previously thought to be the exclusive domain of cable companies.

Since ADSL uses the same telephone facilities as traditional telephone service, it allows use of existing analog modems, fax machines, and telephones while maintaining a high speed data connection at the same time. Moreover, it can be rolled out quickly as a two-way technology. Many cable TV systems are engineered only one way and require significant capital investment to upgrade, as discussed later in this chapter.

8.8   Cable-Based High Speed Telecom

Cable modems are an area of promise for the telecommuter and small office as well, since many homes and small offices have access to cable television facilities. Cable modems are not necessarily new. For example, I personally undertook a project in 1985 to run T1 traffic over local cable television facilities. At that time, we used Scientific Atlanta Broadband Cable Multiplexers which took a 750 kilohertz segment of cable bandwidth and ran a T1 circuit on it. This is, of course, the equivalent of stone knives and bear claws in comparison to what is available or under development today. Over the years other commercial users have attempted the same, but the ability for a *homebound* user to order a data circuit was virtually nonexistent. You had to be a large user, and even then it was difficult because the results varied and horror stories abounded.

In the 1985 example described above, the T1s carried by the cable television franchise had a very good service record for the three years they were utilized. Other subscribers who made similar efforts had more dismal results because the quality of cable systems (and attendant support and troubleshooting systems) varies greatly by company and by franchise area. For example, sometimes a cable company is not staffed to provide 24-hour support. That kind of service is optional for a TV watcher but absolutely necessary for a data service user. Other times the cable systems are only set up for one-way transmission, since TV only goes one way. Two-way amplifiers are an expensive upgrade. Support systems and troubleshooting tools also lag, since in the past, data was ancillary to the cable company's core business and was new to support personnel. These are all still issues. What is changing, however, is that the cable companies are hitting these concerns head on with massive spending to upgrade their networks, add data capabilities, add telephone switches, and bag the lucrative data service customer. They are also filing "COAs," or certificates of operating authority in most states to compete head to head with the local telephone companies. In many states, they are already approved. All this occurs while cable company revenues decrease because of satellite competition. It will be interesting to see how everything shakes out.

8.9   Equipment Your Users Will Need

Sophisticated cable modems are making a comeback today, due in part to regulatory changes that allow cable television companies to get into the telephone business. The concept behind the cable modem is still essentially the same as what we used years ago. You take a specific frequency range in the cable spectrum and run the data service over it. However, cable facilities have a *much* higher bandwidth than copper facilities, so speeds of multiple megabits per second are possible. This is cable's biggest advantage, and one that must be fully exploited.

One popular cable modem today, Motorola's CyberSurfer modem, transmits at 768 Kbps but receives at full Ethernet speeds of 10 Mbps. One of the best uses of cable today is for receiving large files from the Internet. At least that is where the manufacturers of cable modem equipment are putting their money. Cable modems are being deployed with an Ethernet cord on one side that connects to a local area network or personal computer. As far as the computer is concerned, it is hooked directly to the Internet via an Ethernet cable. In fact, it is not even necessary to dial a telephone number. The user enjoys speeds varying from 500 Kbps to 30 Mbps which far surpasses anything else available, at least at the time of this printing. The same facilities, presumably, will be used to deliver television programming, interactive games, and telephone service. Quite a tough package to beat if deployed correctly.

8.10   Another Cable Threat—Direct Satellite Service

Cable is not alone in the high bandwidth business. For example, direct broadcast satellites are one-way devices, just like cable. However, scenarios are already being devised for a low speed data channel to an Internet service provider where an ISDN or standard telephone can act as the "upstream" channel and trigger data downloads of multiple megabits per second using a high capacity satellite link as the downstream channel. The pricing dynamics of this scenario would be awfully tough to beat, since there is no investment for outside plant. The only investment, again, would be in the small satellite dish to go on the roof of the user. These are up, running, and available at the time of this printing.

8.11   Cable Engineering Pitfalls

Typical cable systems serve between 500 and 2,500 homes on one line. The cable carries multiple signals or channels. Most of these would naturally be expected to be devoted to TV but data service would run right alongside. A typical TV channel, used strictly for data, will deliver gigabytes of data to hundreds of individuals with Ethernet quality through-put. However, as users begin to proliferate on these systems, bandwidth and congestion problems can be expected. This has already been experienced in some universities where dormitory buildings are wired for high speed access and the resulting congestion problems cause throughput to the user which is not a whole lot better than a 28.8 Kbps modem. This puts the cable companies in the unusual position of hoping that they will be successful, but not *too* successful. Think about it. The telephone companies have had 120 years to learn how to traffic engineer their networks, yet they still get caught with their pants down when Garth Brooks puts concert tickets on sale, or when other anomalies cause network congestion. There will be a learning curve for cable in this regard as well.

8.12   Standards Concerns

Cable modems, being fairly new technology, differ significantly in their data transmission specifications. Modems from different manufacturers are most often incompatible. This is a concern if a user moves to another city, for example. It would probably be necessary to lease *another* cable modem from that local cable company. Standards are on the way, however. In 1997, nine of the major vendors announced that their products will interconnect with one another. Four providers (AT&T, Intel, Hewlett Packard, and Hybrid Networks) are planning to publish a standard soon.

Think about it, one day you might be tuning in to the "Internet Channel" on your television set! Some problems do stand in the way, however, in deployment of cable modem technology, including financing. As I stated earlier, many cable companies are experiencing severe cash crunches today. At the same time, they are forced to be making major capital investments to provide data and telephone services—all while revenues from their mainline operations are falling. All the more reason to work smarter, and tailor services to cash-rich high profile users.

8.13   Crafting a Strategy—How Cable Should Respond

Whether provided by telephone or cable companies, the networks of tomorrow will require close collaboration between carriers, equipment vendors, and government regulatory authorities. Does your company simply wait for its equipment manufacturers to come in and dictate the direction of technology? If so, it runs the risk of becoming road kill on the information superhighway. True innovators are dabblers and experimenters. They test technology. They stretch their customer's imaginations to the limit. In the process, they leapfrog ordinary technology and their competition, since they invent and get to dabble in the newest, highest margin services first.

We will discuss three specific areas, and hopefully learn how to proactively drive the upcoming revolution in telecommunications. The first is essentially what "everybody else in cable will be doing" namely, raiding the telephone company's markets. There are right and wrong ways to do this, and we will provide you with a few tips. The second is what forward-thinking cable companies will be doing to exploit the "SOHO" (small office/home office) market—one which they are uniquely positioned in many ways to serve. Lastly, we will tell you a little about what big companies need, and how a forward-thinking cable company can exploit this difficult to penetrate but highly profitable market segment by differentiating their services and solving pressing business problems.

The availability of unprecedented amounts of bandwidth in local and wide area networks afforded by cable companies will change the dynamics of how we do everything. How it is packaged, presented, and perceived by the potential customer is paramount in importance. Hopefully this book will assist you in deciding which markets hold the most promise for your particular company, and in crafting your strategy to profit from them.

8.14   Cable Market Opportunities

Figure 8.4 illustrates three distinct opportunities for cable companies. For clarity we have illustrated them as:

- The Two Dollar Window

- The Five Dollar Window

- The Hundred Dollar Window

8.15   Traditional Telephone Service Resale (Two Dollar Window Services)

This is what just about everyone in cable is looking into doing. And why not? At least initially the margins look appealing and it is easy to get into. Resale of residential and business telephone service will become a major part of the revenue stream of many cable companies. Already, some cable companies are in the multitenant sharing business serving major apartment and office buildings. While a cable company is building a name in the business they can capitalize on certain "captive" markets such as apartment or condominium complexes. Renters many times like the "all bills paid" convenience afforded by apartments. Items like electric service are already included in the monthly rent so why not telephone services too? A nominal monetary incentive to an apartment complex owner may even be used to secure a block of customers for not only cable TV services, but local telephone service, long distance, and Internet services as well. Caution should be the operative word however. Already we have heard stories of a few cable companies locking in customers, then denying choices to long distance carriers and engaging in anticompetitive practices by virtue of their "monopoly" position in the apartment building. It can be expected that regulatory agencies will stand rather hard on this kind of behavior when the complaints start coming in, just like in the 0+ operator services business a few years ago.

So what is the best way to exploit these kinds of markets? Unquestionably for the present it is through the purchase of unbundled components from the local telephone company. Your cable company will have to file for certification in the states it wishes to serve (if it has not already) and once approved, it will negotiate an interconnection agreement with the local telephone company in the service area desired. This company is now called an ILEC, or incumbent local exchange company. Your company will technically become a CLEC, or competitive local exchange company. A CLEC can buy unbundled components from the ILEC to build and market whatever kinds of services it desires. For example, you can lease a copper "loop" from the ILEC and connect it to your own switch. Your company can also lease T1 services, interoffice switching, billing and collection services, operator services, and myriad other services under terms and conditions negotiated with the ILEC in

"The $2 Window"

Traditional Telephone Resale, Using Unbundled Components

"The $5 Window"

Marketing Internet Access and Telecommuting Solutions to the Small Office / Home Office (SOHO) Environment

"The $100 Window"

Upgrade of Fortune Level Clients to SONET Technology, to Include Client-Server Solutions and Network Management Services

Figure 8.4: Three new profit opportunities for cable companies.

the interconnection agreement. Specifics of exactly how to do this are far too complicated to be included in this book. Suffice it to say however, if your company is astute enough to assemble unbundled telephone components and combine them with other services, the margins are good and prospects for profit exceptional.

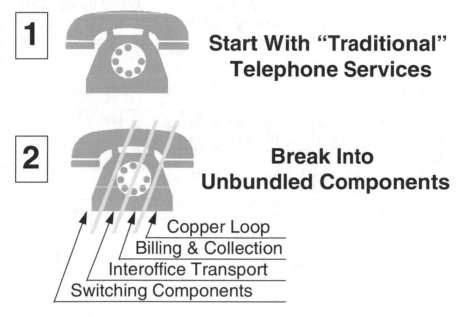

PROJECTION
The average resale rate being adopted by state regulatory commissions is around 20 percent. Companies astute to assemble *unbundled* components purchased from the telco can expect much higher margins, at least initially.

Figure 8.5: The $2 window, traditional resale.

8.16   Enhanced SOHO Services (Five Dollar Window Services)

Next to consider are more emerging technology–based services such as Internet access and advanced telecommuting services for the small office/home office (SOHO) market. In many ways this is one of the areas where cable is positioned to take the lead.

There are areas in virtually every metropolitan service area where people have the demographic profile in terms of income, occupation, and technical savvy to avail themselves of high speed data services to the home or small office. Internet access in particularly lucrative at the present time. Anyone who has every surfed the Internet at 14.4 Kbps knows that the speed of traditional modems, even the newest 56 Kbps variety (when they run at that speed) is woefully inadequate. ISDN lines provide some relief, but even 128 Kbps is underpowered for large downloads, graphics, video, and multimedia. That is where cable modems come in, however as pointed out earlier, this is easier said than done for the cable company. Large capital investments must be made to upgrade the network to two-way traffic for starters. Or is there a way to sidestep this expense at least initially? There is. Read on.

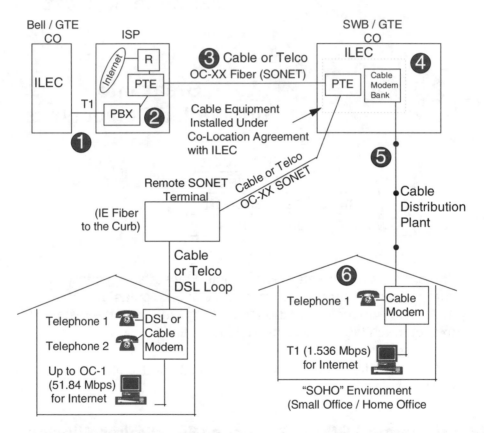

Figure 8.6: The $5 window, new services for the SOHO market.

8.17   Competing on a Budget

Think back for a second to that certificate of operating authority your company filed for. Remember, it allows companies like yours to buy unbundled components from an ILEC for resale. Just suppose you combined this capability to buy what you need from the telco with what can be provided in your own network. Consider the example in Figure 8.6. The key here is to be sure your certification does a few things, namely:

1. Provide for *co-location* of your equipment in the ILEC's central office. This will give you the meeting point you need to connect your equipment to theirs.

2. Make sure your certificate of operation in the state where you are doing business allows you to be a *facilities-based* carrier. That way you can use your network whenever possible to save money, but round it out with the telephone company's network where it makes economic sense.

Figure 8.6 shows but one of many possible solutions, in this case to provide high speed Internet access to an affluent subdivision with high demand for such services.

8.18   The "Fortune" Level Customer ($100 Window)

This class of customer might just represent the most demanding, highest cost, least loyal, and most highly competitive market segment your company will tackle. It might also just be the most profitable. Large corporations have insatiable appetites for telecommunications services of all types. They historically play hardball with all service providers, shop the market thoroughly, and take a sharp pencil into negotiations. Nonetheless, in many cases they possess the intellectual savvy to understand even your most technologically advanced offerings and have the muscle and capital to help you develop them—allowing you to sell the process elsewhere in years to come. Having a few of these bellwether users under your belt is therefore very useful indeed. To even have a chance in this market, however, the cable company needs to understand the unique dynamics of these organizations, and be in tune with what they care about. They must also be prepared for a long sales cycle, and be prepared to craft honest-to-Gosh business solutions. Here are a few tips.

8.18.1   Fortune Users Have Insatiable Demands for Capacity

Users want to share larger and larger files, taking a greater and greater toll on network capacity. They need higher capacity services and T1 is no longer enough. Cable is technologically capable of providing 10 and 16 Mbps services—native LAN speeds. These are much more appealing to this class of user and will sell.

8.18.2   Fortune Users Need Help Desk Support

Help desk support is a big deal. Any value-added services that the user can get from his supplier are therefore a big selling point. As a minimum, do not expect this potential user to take on any additional overhead in this area just to accommodate your company, they will show you the door.

8.18.3 *Fortune Users Need Data System Backups and Archival*

Could your company provide this user enough capacity to do things like *on-line vaulting*? That is where the user has enough affordable capacity (T3 level or better) to connect to a remote location and do transaction backups in real time off site. It is a very salable commodity to banks, brokerage companies, and other firms where last night's tape backup is obsolete at 9:05 a.m.

8.18.4 *Fortune Users Need Network Management*

This may be *the* most important consideration. Before opting for an "unproven" carrier, the Fortune level client will want to see a Johnson Space Center level of network command and control. There are sound business reasons for improving your network management and control capabilities. This includes providing (or selling) high level network management services to client-server customers essentially "black boxing themselves to death." Many information executives and end users alike are having second thoughts regarding management of technology with sentiments such as:

- *Can we really afford to chase technology?*

- *Do we really want to chase technology?*

- *Wouldn't it be better to focus on our core business instead?*

Therefore, anything we can do to take the focus off maintenance, repair, and troubleshooting and allow the end user to concentrate on core business issues (the things that bring in money for the company—like selling things!) will be an exercise in enhancing efficiency and the bottom line. The message to the end user: "*We'll* manage the network, *you* concentrate on building Chryslers." It is a powerful pitch if you can really pull it off.

8.19   Selling Network Management to Fortune Users

Does your network management system provide your organization with all of the following capabilities?

1. Performance management

2. Accounting and chargeout management

3. Change control

4. Alarm and problem resolution

5. Audit and inventory of asset management capabilities

6. Capacity planning for future needs

You will need these kinds of capabilities to be up to par with that bank, factory, brokerage house, or other demanding user. You will also need to refine your capabilities in these areas as your own company's network becomes more complex due to the new directions we have outlined.

If you bat 1,000 on all of these issues, then your company has a chance to crack these potentially lucrative markets. Then the fun starts.

The additional capacity that a well-designed cable company can bring to this class of user will foster new services, some of which will permanently alter the way that user does business—and the way you do business. The things you learn by dealing with these new classes of customers can be replicated elsewhere. They can also be licensed, shrink-wrapped, and resold to less forward-thinking companies (after you have moved onward to bigger and better things), providing new sources of revenue.

8.20 "Hybrid" Solutions

Cable system owners facing major capital expense to upgrade their systems often utilize a hybrid system as shown in Figure 8.7. The user dials up his upstream path, while signaling equipment to transmit Web data via a high speed cable channel. Similarly, the same technique is used with small KU-band satellite dishes like Direct-TV.

8.21 Summary

With regard to cable-based data connectivity, a few hurdles exist before high speed connections are commonly available from cable television companies. Notwithstanding the issues of capitalization, standards issues will also have to be addressed to ensure compatibility. Another problem is centered around training. Just as the phone companies are struggling with providing ISDN and advanced data services, cable operators also must come up the learning curve regarding providing these sophisticated data services. Lastly, there are considerable technical support hurdles.

While it will be at least months and perhaps years before cable modems are available in your local BizMart, this is an exciting technology and should be closely watched over the coming years. As an ever-increasing hunger for bandwidth affects heavy Internet users, you can expect access to the Net to become a major catalyst in driving the success of exciting high bandwidth solutions such as these.

ADSL and VDSL face similar start-up problems, combined with a natural trepidation on the part of the local telephone companies to embrace an expensive technology that may easily become obsolete due to competing technologies. ISDN is a safe bet for now since it is relatively inexpensive for the local carriers to deploy. ADSL and VDSL will cause a bit

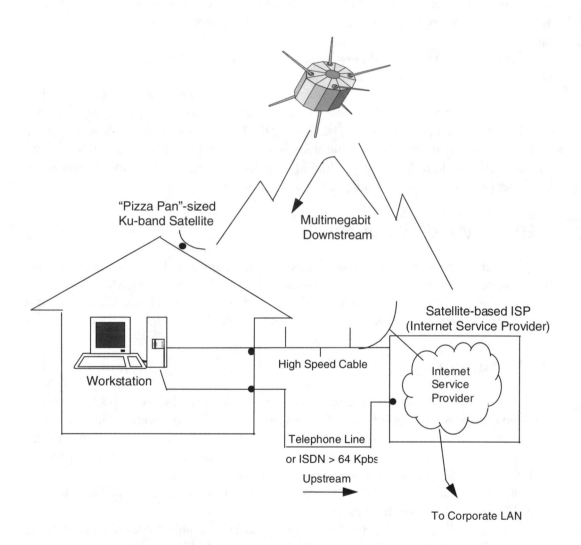

Figure 8.7: Direct satellite broadcast—Internet access.

more heartburn for these carriers, since they can be easily eclipsed by cable modems and other media. Look for a slow, cautious, but steady rollout of these services. Remember, ADSL and VDSL are copper based and also relatively inexpensive to deploy and can be presumably embraced by the carriers for that reason.

Direct Satellite Broadcast systems are yet another wild card. It is quite feasible to use a landline to an ISP (Internet Service Provider) to initiate requests for downloads beamed in by satellite at speeds eclipsing even the fastest cable modems. It is simple technology, so simple in fact that systems are already on display in a few retail organizations and satellite TV shops. But these systems will not easily provide telephone service and in fact require an outboard dial tone line to work. Nonetheless, they cannot be ruled out as effective market entrants.

In short, even though we have only scratched the surface, we have hopefully illustrated what an exciting and dynamic era we are entering with regard to deployment of new services, befitting the start of the new millennium. Enjoy the ride!

9

Exploiting Telecom Reform in Network Procurements

9.1   New Technologies = New Advantages

Did you hear? It seems the U.S. is number one again. According to former FCC Chairman Reed Hundt, quoted recently at a major trade show, "Competitive, cheap, efficient networks are the reason the U.S. is now number one in productivity." He went on to add that "private market forces build networks if they work competitively."

The role of government, in this case the Federal Communications Commission, is to prevent unfair competition and protect the public interest, something the private sector cannot always do. *Innovations, however, must come from us.* The "Information Highway" must be married with real-life business solutions to produce increased growth and national competitiveness. When a technologist preaches the gospel of growth and jobs, it is nothing short of surprising what he or she can accomplish with regulators.

So how will it happen? One school of thought has the government driving progress, in much the same fashion as it was the catalyst behind construction of the interstate highway system. Another says, "Deregulate us and it will happen!" In fact, for real progress to be made, we must look between these two extremes. It is here that we find the world of *partnerships.*

High-profile corporate users also help communications service providers clear regulatory hurdles with skeptical commissioners, while at the same time, ushering everyone involved in the partnership into the Information Age. No serious commissioner will say he does not agree with jobs and growth, provided of course that the public interest is protected. Additionally, for the first time, regulators can directly facilitate innovations in communications for their state or jurisdiction as actual working partners in the process. In order to accomplish all of this, and position your company to become a technological superstar, you can take a series of coordinated and preplanned steps now.

Many federal and state regulators are trying to foster innovative partnerships with the corporate community, particularly in the area of advanced technology. Indeed, this was the underlying sentiment in the Telecommunications Reform Act of 1996, the most important piece of new legislation in our field in years. This means that if you are so inclined, the regulatory vehicles exist for your company to have the opportunity to deploy more than "standard telephone company offerings" and truly "step out of the box" with advanced networking solutions.

This will of course require a collaborative effort between telecommunications service providers, users, and government. If business drives the process, we can be quite sure that products will be developed that will be usable to business, and that will help foster productivity and growth. For this reason they are equally desirable to the carriers, as an opportunity to break out of zero-sum-gain markets and market radically advanced applications to other users. Lastly, regulators can expect to be enamored with any concept that creates growth, jobs, and U.S. competitive advantages.

To make this dream a reality, we must sharpen both our technical and our political skills and become activists in the regulatory process itself. How does one influence a potential partner to build a custom network? How does one influence a commissioner to approve the deal? Before we jump headfirst into technology (and believe me there will be plenty of that later) let's look into the issue of what having tomorrow's network today might really mean to a company.

9.2   Comparing Physical and Electronic Distribution Mediums

There are not many other people around who appreciate a superhighway more than I do. Some years back I lived in Tokyo, Japan, population 11 million at that time, give or take. The thing I missed most about the USA was our expressways.

If you can imagine, Japan today has one car for every 32 feet of paved surface which includes driveways, alleys, roads, whatever. It is not unusual at all for a 30-mile ride to take six or seven hours if you go by car. So what is the result? As one would imagine, people tend to avoid the roads whenever they can and opt instead to take a train, ride a bicycle, or walk.

I have a lot of respect for the Japanese and the powerhouse economy they have built. But their internal system for physical distribution of people and product is abysmal. If however, someone was able to build a system of American-style freeways six or eight lanes wide, it would change the dynamics of how they do everything over there, and their economy would soar. I mean, think of how infinitely simpler it would be to move things around!

Today, however, it is not so much physical product that needs to be distributed as information. The point here is not to draw the same parallels everyone likes to draw between information highways and physical ones. The point is to illustrate that the upgrades taking place today in the U.S.'s communications infrastructure will be no less dramatic in their effect than building a 12-lane superhighway from Hokkaido to Okinawa.

All of us in *this* country, currently avoiding the information transport system by doing the electronic equivalent of riding the train, taking a bicycle, or walking, will be literally sucked in by the changes. It will change the way we do business forever.

Even a 12-lane highway is of little use if a company has never thought of using trucks before, or if they had never even thought of distributing their product. Think about going to the doctor for an X-ray. Only today are people in this field even thinking about distributing their product. The network is maturing. Soon it will be more economical to have a doctor in Philadelphia review an X-ray of a patient in Dallas, simply because the network will be cheap and available. That is part of the reason why we build cars in Flint, Michigan, and distribute them rather than have manufacturing plants in every backwater town. It is also the reason why distributed processing's days are numbered. It makes no sense to have a full-blown server or mainframe in every backwater town when the ultimate unifying factor, a 21st century communications infrastructure, is right around the corner.

In a hazy networking future, a few things seem certain:

- Radical changes regarding advanced telecommunications are coming in the near future.

- They will completely redefine how we do business.

- Changes will be monitored by government regulatory authorities. (Can you imagine anything this powerful and far-reaching not being regulated?)

- We must learn not only technology, but also business processes and regulatory skills.

- We will be the ones who drive this process.

- We must be able to articulate business needs and technological requirements to what may be nontechnical governmental commissioners.

- The potential payback for those who do these things is awesome.

The result is the ability to do a "custom network deal" by crafting these emerging network services into a strategic tool, not easily available to your less-technology-savvy competitors—a custom tariff if you will, for the network of tomorrow.

9.3   Communicating the Vision

Custom network assemblies and tariffs are not new. Any company of size probably has an AT&T Tariff 12 arrangement for long distance service. But what we are talking about goes far beyond saving a nickel or two on long distance. We want to marry advanced business requirements with enabling technology. Here is how we do it.

Consider Figure 9.1. In this scenario, a nontechnical person in the core business has a brilliant idea. Remember though, he is nontechnical. This means he has to explain his idea in the best way he can to the technology planners, shown in the center of the figure. Also remember that these people are up to their necks in the technology itself, which leaves little time to self-actualize about what the business is really supposed to be doing. The technology planners, in turn, relay the information to an executive or financial person, who (at best) understands about 25 percent of the original idea represented, due to the "translation" process. So how is technology usually implemented? An executive in the core business makes the classic "end run" to an executive in the financial arena and screams, "We're dying out here, we need it *now*!" Then sometimes it happens. It is not a very graceful way, however, of implementing technology. Now, suppose the original brilliant idea involves technology not available off the shelf. It is a true 21st century network application. It is technically possible, but must be custom built.

The application is the ability to upload a 4-gigabit chip diagram to a supercomputer and receive an immediate response. The potential payoff? Cutting the design time of a custom chip from six months to six weeks. In the brutally competitive semiconductor industry— potentially hundreds of millions of dollars. Sounds like a pretty ripe prize to go for, doesn't it? (Incidentally, this is based on an actual company.)

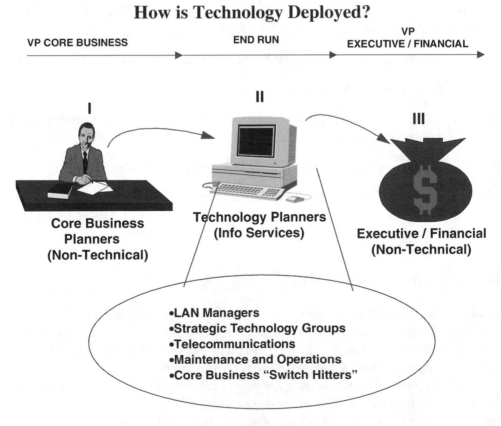

Figure 9.1: Communicating business vision.

To do this, we will need a 622 Mbps SONET ring, with on-site ATM hubs. It will be installed on a custom basis through a joint partnership with one or more service providers. Since it will fast become a revenue impacting system, we need to build in bulletproof protective systems as well. This will probably involve tapping into network diagnostic data directly from the carrier, and interfacing a Johnson Space Center level of network control.

Since the deal will involve regulated carriers, however, the Federal Communications Commission, state public utility commissions, or both, may also get involved. Even in cases where the carriers have flexibility, it is important to know the rules. After all, we still cannot generally just pick up the phone and order an OC48. Someone in your state capitol will want to know the whys and hows.

9.4   Selling the Concept Internally

Lastly, we will have to sell the boss on making a large capital investment; articulate the business vision to myriad departments, both internal and external; file paperwork required at the state and federal level to register as a lobbyist; and so on. Sound complicated? It is. But think of the potential payoff. And if it is difficult and complicated to initiate, it is all the more difficult for your competitors to copy. That means your company enjoys the strategic advantage longer before others jump in. And what is in it for the carrier? They stand to benefit most of all. The application you develop with them can be licensed, shrink-wrapped, and sold again to another customer (after you have moved on to something even better I should add.) That sure puts the carrier in a different class from those peddling long distance service at yard sales and swap meets! And incidentally, the more lucrative, tantalizing, and broadly adaptable the application is, the more accommodating you can expect the networking partner to be. This will be very important when you get down to the point of up-front concessions, who will buy what equipment, and what it will ultimately cost.

By now you have realized that we are describing a radical new process demanding many new skills. Since whole books can be written on just a single one of these topics, we have selected eight specific strategies you can use to bring your custom network to reality. While they do not represent everything, they will get you well on your way!

9.5   Kicking Things Off

Remember, to keep improvements and capital expenditures consistent with your company's mission, effective linkage to its core business is essential. A detailed methodology for graceful integration of technology is shown below and in Figure 9.2, embodying the eight specific strategy points I mentioned in Chapter 6.

1. Conduct detailed business process interviews.

2. Develop specific "feature packages" of services based on business need.

3. Divide packages into "existing" and "emerging" technologies.

4. Negotiate price on existing technologies in traditional fashion.

5. Develop "short list" of potential emerging technology partners.

6. Establish executive liaison and joint venture with selected partner(s).

7. Negotiate price based on LRIC.

8. Implement new technologies, share access to labs.

9.6   Doing a Custom Tariff—Learning the Rules

Every state utility commission has its own rules. For example, some will tell you whether you will need to have a lawyer certified in the state to represent you, or how many copies of testimony you will have to file. At an early stage of the planning process, you will have to learn these rules and incorporate them into your attack plan. Hopefully, you have maintained contact with the commission staff after the original meeting with the commissioners themselves, and have a good relationship with them. They can probably help you with some of the subtleties.

9.6.1   Learn the Language

Administrative lawyers, or others who follow regulatory affairs, like to segregate the types of cases the commissioners hear into two types:

1. Rulemaking/Legislative

2. Contested/Judicial

In a rulemaking case, the commission is adopting a statement of policy or change in rules that will affect a specific class of utility under its jurisdiction in the future. A contested case has the commission usually making decisions based on past actions, and determines the regulatory response based on the policy it makes, which in turn is based on the specific facts of the case.

Any questions so far? In reality the two methods overlap and the courts usually give commissioners a wide degree of latitude in deciding which method to use in establishing policy. Unless your company already has an expert on the operations of your particular commission, you will probably need to retain outside help at this juncture.

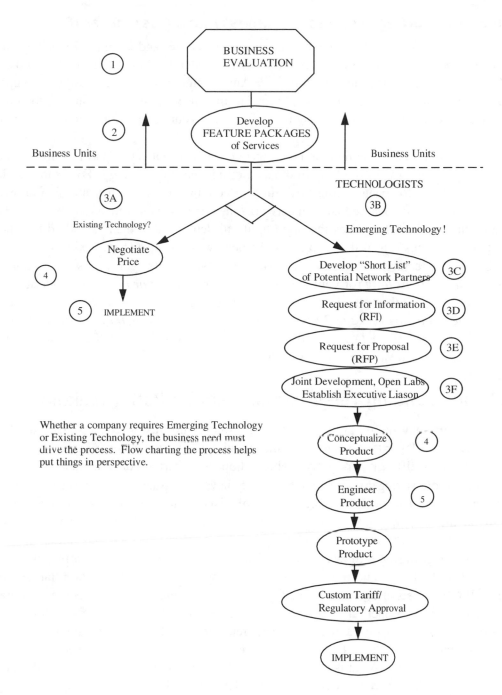

Figure 9.2: Graceful integration of emerging technology. Regulatory approval will be a key component.

9.6.2   Converting "Feature Packages" into a Custom Tariff

A tariff is an agreement between a communications carrier and a regulatory agency such as a state public utility commission (PUC) or the Federal Communications Commission (FCC). Tariffs command the weight of law. Any communications services your company orders are probably tariffed services. This means that the same services available to you are also available to anyone else desiring them, including your competitors. Here is where the project starts to get fun.

Many federal and state regulators are trying to foster innovative partnerships with the corporate community, particularly in the area of advanced technology. For example, they may be inclined to grant pricing flexibility on certain services. They are also allowing companies to build advanced custom networks on a one-time custom assembly basis. This means that your company has the opportunity to deploy more than "standard telephone company offerings" and truly "step out of the box" with advanced networking solutions.

Using the feature packages as the basis for your efforts, you can secure a *custom telecommunications tariff* for your company, for long distance *or local telephone services.* That is why you go through all the trouble. *When regulators can see all the steps—Business Requirement, Feature Package, Technology Partner, Custom Tariff, they see the reasons you need a special service and will be more inclined to grant your request.* Then your company has a competitive advantage that no one else does!

9.7   Influencing Telecommunications Policy Makers

Many readers may not be aware, but I am personally not just a technologist, but also a former elected Mayor for a suburban Dallas municipality. This presents quite a contrast in jobs, and a very different perspective, which I hope to impart to you, the reader.

Like the old saying goes, anyone who loves laws or sausage should never watch either one being made. I'll vouch for that! As technologists, we look for normal, rational ways of doing business. However, the world of public policy is very different and, on the surface at least, not always rational to an outsider.

As a former mayor, I know first hand what it is like to consider a case, armed with only one third of the facts—and those "facts" from lobbyists! I also am well familiarized with the feeling one gets after making a bad decision and having 150 angry people to contend with jamming the council chambers the following week. I know the value of an unbiased expert, although in the public realm these are few and far between. I empathize with utility commissioners completely. I mean, I can sure see things from their perspective. Do you believe the lobbyists? What is the hidden agenda? Who will scream after what decision? Will we be challenged in the courts? It's a wonder that in this environment anything can get done!

There are a wide variety of influences that will apply to a utility commissioner. These might include congress, consumer groups, the governor, the legislature, utilities, commission staff members, the courts, and even the commissioner's family. Also realize that no commissioner is an expert in everything, indeed, he or she may not be conversant in telecommunications at all. Their background may be in economics, politics, law, or even being a housewife. It is kind of like being the captain of the track, football, baseball, and basketball teams all at the same time—possible, but only if you surround yourself with experts to advise you. That is why they have both technical and legal staffs. It is also why they need input from you, the technologist.

At the risk of offending people, we cannot continue to allow lobbyists and regulatory attorneys with hidden agendas to set our national telecommunications policy. We, the people who *build things*, must take our rightful place in the process too. That is what creates jobs. While not the primary objective, there are some very good cost implications as well. Look at the past history of the rate setting process. Who has traditionally paid most, while at the same time having the least input into the regulatory process? Why *you* of course! And companies build and create jobs—they should be first in priority, but instead are last. As I stated, the process is not always logical to the outsider. That is getting ready to change—read on.

Consider the plight of today's regulated communications carrier. For these companies, a high profile corporate user can help clear many of the regulatory hurdles with skeptical commissioners. At the same time, everyone involved in the cooperative arrangement prospers—your company, your carrier, the commissioners, and the general public. That's right, even the general public prospers because the proposals you make (if you do the job right) will create jobs. No serious commissioner will say he does not agree with jobs and growth, provided of course that the public interest is protected. Additionally, for the first time, regulators can directly facilitate innovations in communications for their state or jurisdiction, as actual working partners in the process. This makes their jobs exciting and useful as opposed to something perceived as traditionally adversarial.

If *business* drives the process, we can be quite sure products will be developed which will be usable to business, and which will help foster productivity. These might include some glamorous, high profile projects such as:

- On-line vaulting, archiving, and imaging

- Realistic simulation of designs prior to production

- Concurrent engineering and collaborative work groups

- Distance learning and distance medicine

If you are a carrier, ask yourself this question: Would you want to work on exciting projects that build national competitiveness, or concentrate on ways to sell 10 more minutes of long distance service in a box of cereal? If you are a public utility commissioner or staff member, what would you like to point to as a result of your tenure? The key is to get everyone pulling together, as illustrated in the following methodology.

9.8   First, Talk to Each Other!

This is not as crazy as it sounds. As Mayor, I met with lots of people, some who were commendable and others I would like to forget. Technologists who were unbiased in areas of concern to me, however, were always welcome and got appointments.

Take a look at how the world looks to a nontechnical or semitechnical commissioner. It is not exactly like they can corner a staff member in the hallway and scream, "Hey you, have you *innovated* today!?" It is not like they are a bunch of dummies at a utility commission, quite the contrary in fact. But, they are not in the trenches of corporate America and can often be at a loss as to what exactly "we" all need. For this reason, you as a technologist can be somewhat of an anomaly to them in a very positive sense. You are their window to the outside. The genesis of the meeting should be: "I have no agenda now, but I may be back in six months with something really unusual. If this happens, I ask that you keep an open mind." Then open up. They want to hear what you are doing.

Needless to say there are lots of subtleties involved when meeting with a significant public official. Perhaps your company's legal department will want to get involved, or maybe your firm has an established regulatory affairs staff. Even so, do not make the mistake of going in with your company's lawyer or someone who will make the commissioner uneasy. Public officials know you do not know all the rules and tend to be forgiving individuals. Besides, your unadulterated opinions are of most importance to them. The important thing is establishing dialogue and rapport.

9.9   A Little Bit of History

Do not worry—this history will not be like when you were in the eighth grade and had to remember what happened in 1066, or why the *Magna Carta* is important. The Telecommunications Act of 1996 was literally revolutionary, so that much of the history of the past will not need to be known by telephone gurus in the future.

The basics: Once there were many companies offering local telephone service. Out of those many companies, one grew and became AT&T. AT&T was *the* phone company for much of the nation, owning everything relating to telephones—from the instruments and the wires and the switches to the phonebooks themselves. The Department of Justice thought the company was too big, and filed an antitrust case against it. The case was settled by an agreement that split AT&T into one long distance company and seven regional Bell operating companies (RBOCs). The case was under the jurisdiction of the federal district court in Washington, D.C., which decreed that the RBOCs could only offer service within limited service territories, called local access transport areas, or LATAs. Within a couple of years, the RBOCs were chafing under this restriction, and sought permission from the district court to offer interLATA service. Judge Harold Greene turned the RBOCs down, so they started lobbying Congress for the ability to get into the interLATA market. See Figure 9.3.

- ◆ **AT&T**
- ◆ **The Baby Bells**
- ◆ **The IXCs**
- ◆ **The FCC**
- ◆ **Congress**
- ◆ **Harold Green and the Courts**
- ◆ **State Regulatory Commissions**
- ◆ **Consumer Groups**
- ◆ **Competitive Local Exchange Carriers (CLECS)**
- ◆ **Cable Companies**
- ◆ **Cellular Companies**

Large Institutional Users – <u>YOU</u>

Figure 9.3: The players today.

Meanwhile, the companies that were in the interLATA market—interexchange carriers (IXCs)—were having to pay the BOCs money to carry telephone calls the last few miles from the IXC switch to the home. These "access charges" were deliberately designed to be high, so that local telephone rates would not be driven upwards. Some folks got the bright idea of building their *own* lines within the territories of the BOCs, and avoiding these high access charges. These folks, called "competitive access providers" (CAPs) found that there were numerous legal and policy barriers on the state level, and they had to go through many fights in order to compete against the BOCs. Thus, they started lobbying Congress for the ability to get into the local market.

At the same time that the BOCs and IXCs were fighting with each other, the cable companies were finding out that they were building communications poles and wires, and had the idea that they could start doing something *more* with those poles and wires—sell telephone service. However, there were restrictions in the federal law that prohibited *them* from providing telephone service. Thus, they started lobbying Congress for the ability to get into local *and* long distance.

All of these forces came together and lobbied Congress for years. Finally, the forces reached a compromise—the BOCs could get into long distance, but only after they took steps to open the local market to competition. Any state rules or statutes that prohibited that competition would be preempted. The cable companies were given the ability to get into telephone, and the telephone companies were given the ability to get into cable. Once the pieces fell into place, the legislation moved very quickly and became the Telecommunications Act of 1996 (see Figure 9.4). Under the Act, there are several important levels of regulation on the federal and state level.

◆ Telecommunications Reform Act of 1996

◆ Initiatives in the Individual States

Figure 9.4: Recent changes.

9.10   Federal Law and the FCC

The chief policymaking agency in the federal government is the Federal Communications Commission (FCC). It is composed of five members, appointed by the President with the advice and consent of the Senate. It has the principal responsibility for interpreting the Federal Telecommunications Act. At the time I am writing this chapter, the Federal Court of Appeals for the Eighth Circuit has ruled that the FCC did not have the power to set rules that described how the state public utility commissions (PUCs) set prices for resale and interconnection under the Act. The Eighth Circuit's opinion has been accepted for review by the United States Supreme Court. By the time you are *reading* this chapter, the Supreme Court may have interpreted the Act in relation to the pricing issue. I anticipate that the Supreme Court will find that the FCC had more power than the Eighth Circuit thought it did—but agree that the FCC did not have as much power as the *FCC* thought it did.

The FCC had numerous other tasks assigned to it under the Act. You can learn a great deal about the FCC's actions by using the Internet to go to www.fcc.gov/telecom.html. You will see lists of dockets. A quick help—"NOI" stands for "Notice of Inquiry" which is the first step in most rulemaking. In an NOI, the commission asks questions to the industry about areas in which they are considering rules. An "NPRM" (Notice of Proposed Rulemaking) is the next step in the process, where the Commission actually proposes some draft rules for comment. Finally, the Commission will issue an "Opinion and Order," in which it finally decides the language of the rule. The FCC frequently reviews its orders upon reconsideration, so there may be several "Opinions and Orders" in a given docket.

9.11   State Law and the PUCs

Each of the states also has regulatory commissions that have jurisdiction over the telephone companies that operate in their state. You can find Internet links to most of them through www.puc.state.tx.us. NARUC is the "National Association of Regulatory Utility Commissioners" and is the national organization that represents the state commissions.

Every state commission is different from its neighbors. The names may be different— Public Service Commission, Public Utility Commission, Energy and Telecommunications Commission, Corporation Commission. (In this book, we will refer to them as PUCs.) The membership can range from three to seven, can be appointed or elected, and can serve from two to seven years. In making telecommunications policy, they generally use either rulemaking or contested cases.

Under the Federal Act, the states are given the primary responsibility for "arbitrating" disagreements between incumbent local exchange companies (ILECs) and competitive local exchange companies (CLECs). They also have the responsibility for crafting universal service rules and regulations. Under their state law, they have the responsibility of making sure that rates are just and reasonable, and most require a company to be "certified" before it is allowed to provide service in their state.

9.12   Local Law and the Cities

Many regulatory practitioners forget about the fact that municipalities also have an important role to play in telecommunications regulation. Traditionally, the city granted a "franchise" to the local telephone company for use of the roads and facilities of the city. That franchise was similar to a tax, and provided revenues to the city coffers. Now that new entrants want to take revenues away from the incumbent telephone companies, the cities want to make sure that they maintain at least the same level of revenues as they did prior to the Act. They are also concerned about the unfettered use of local rights-of-way and roads. Under the Act, the city may set "nondiscriminatory" fees for franchises.

9.13   Regulation Varies by Technology

You have probably read about various "voice over the Internet" providers, where with the right equipment you can make telephone calls from your computer. At this time, providers of such services are not required to register with state or federal commissions, and do not have to pay "access charges" to local exchange carriers. The Clinton Administration is opposed to taxation of Internet service providers. The voice technology is improving at an exponential rate.

If you are an entrepreneur and think there is an opportunity here, this difference in regulation is a good thing. If you are getting into some *other* branch of the telecom business, the special treatment of voice technology seems unfair. The Internet network is not designed to handle the kinds of calls that may originate because of the voice technology, and the telephone companies that own the backbone of the Net need some method of recouping the costs that such technology imposes on their systems. I anticipate that the situation will change so that the voice technology has to pay *something*. Then there will be opportunities, because we will need to decide how to capture what such companies owe, create systems to bill that amount, hire employees to handle disputes over the amount. Each solution brings with it its own opportunities.

Speaking of differences in regulation—the Internet itself is remarkably free of regulation on the state or federal level. I personally think that part of the reason is that the Internet has grown so quickly, legislators and regulators have not been able to respond to establish any level of regulation. This is one case where technology *is* moving faster than regulation. However, I have full confidence that regulators will find a way to start governing prices and service quality in my lifetime.

Similarly, cable companies are developing methods of carrying telephone calls. For the most part, they are submitting to regulation as CLECs. However, effective regulation of a company that is providing two products at once, when there are two different regulators, is *very* difficult. I anticipate that we will have another round of attempts to get cable companies to be regulated more like traditional telephone companies in the next few years.

9.14   The Federal Telecommunications Act of 1996—A Lawyer's Perspective

Note: The following sections were contributed by Eddie M. Pope, Esq. Eddie is General Legal Counsel for our company and former Chief of Staff to the Chairman, Texas Public Utility Commission. I hope you enjoy his perspectives as much as I do!

On February 8, 1996, President Bill Clinton signed the Federal Telecommunications Act of 1996. It was an act full of great promise of competition among major players to provide better services at lower costs. It would give parents the ability to control what their children watched through the use of the "V" chip. It promised to cut down the amount of indecent material on the Internet.

While there are many new providers certified to provide competitive local service, the BOCs still retain huge market shares in virtually every category of service. The "V" chip is still under development. The indecency provisions of the Act were struck down by the Supreme Court. I am sure that many people see the Act as a major failure.

9.15 There Was *Supposed* to Be a War!

All of the parties who favored passage of the Telecommunications Act oversold the impact of the Act on American commerce. Before the Act passed, IXCs convinced Congress that the only thing keeping them from invading the local telephone market was a bunch of backward state laws and regulations. Once the local exchange companies were forced to open their markets to outsiders, big companies like AT&T and MCI would go swooping in to capture numerous customers and provide them with awesome service at lower prices.

The cable interests also claimed that they were ready to bring competition to the home with much more bandwidth and better service. I recall attending a hearing in 1995 where the President of Time Warner Communications promised that they would have every home in Austin wired for telephone within one year of passage of the Texas act. Providers of PCS services also said that they would be competing against the local telephone companies within a year or two—if only the law was changed in their favor.

Much the same song was sung by the BOCs and GTE, who explained that they would be able to offer "one stop shopping" for local and long distance service, and would force the price of long distance calls down once they were allowed to compete. They also said that they would start providing cable service, and drive down the cost of cable TV, while offering better Internet service in the process.

Politicians were convinced that once the Act was passed, we would see an "Oklahoma Land Rush" of major companies competing against each other. They were worried that customers would have so many choices, they would become confused. Telecommunications was going to become a growth industry with new products and services that could only be guessed at in 1995 and early 1996.

9.16 The Act Is a Great Success

The Act *itself* is the reason why we did not see the great rush of competition. When I first read the telephone portions of the Act, my initial conclusion was that this should have been titled the "Lawyers' and Consultants' Full Employment Act of 1996." Page after page is filled with generalities, and simple provisions that would have made the Act more workable were left out. Let me give an example. The Act allows parties with disputes about their interconnection agreements to go to state PUCs for arbitration. It is easy to predict that, with millions of dollars on the line, there will still be disagreements about the decisions made by those state regulators. So, we can expect these state PUC decisions to be appealed. The Act recognizes that such appeals will occur, and says that appeals of the arbitration orders *must* go to federal district court. Unfortunately, it does not say *which* federal court, so it is possible for each party to bring an appeal in a different district. Further, the Act does not tell the reviewing court what standard to use. If the appeal was from a *federal* agency, the law is clear—the court gives a great deal of deference to the agency interpretation of the law, and only overturns the commission's ruling if it was based on an erroneous reading of the law or is contrary to the substantial evidence of the case. Unfortunately, the Act does not say that these are the same standards that apply to appeals from *state* agencies. As a lawyer, I can tell you that there is a reasonable argument that the courts should use a different standard, and should review the law and the facts completely on their own, and make the right decision without paying any attention to what the state agency has done.

From a lawyer's perspective, this ambiguity is wonderful. We lawyers can argue for years in several different courts about stuff like this. We can hope that different courts will rule different ways—that way we can appeal the case to a higher court. Eventually, the Supreme Court will answer the question—but it may be several years from now before the right cases get to the Court. Meanwhile, we lawyers can get rich fighting over the ambiguities.

Of course, from a businessman's point of view, this delay is intolerable. How can anyone expect you to build a sensible business plan if you do not know what prices your main supplier will charge you? This is only one example, and one that will be resolved fairly quickly. There are *other* disputes buried in the Act that will probably take a decade to resolve—and until they are resolved, you cannot really know whether you will have a business that makes a profit or not. Let me give you one more example of why this is the Lawyers' and Consultants' Full Employment Act. Under the Act, the state commission is to set the price of an arbitrated interconnection provision or unbundled network element "based on the cost (determined without reference to a rate of return or other rate-based proceeding) of providing the interconnection or network element (whichever is applicable)." In other words, state commissions are to establish the price of the service based on cost, but may not use the methods that such commissions have used for years to set costs. So PUCs had to come up with ways to *set* costs that were different from the ones that they knew. These new methods were full of items in dispute, on which there was no body of precedent. Different consultants testified that different approaches were the only correct way of applying the law. Different commissions adopted those different approaches—and, of course, got different results. Under *all* of those approaches, one of the parties is a "winner" and the other is a "loser." Since these cases involve parties who have million-dollar legal budgets, we know that there will be appeals over whatever methods the PUCs do use. Some of those cases will be remanded to the commissions, where the lawyers and consultants can earn more money trying to interpret what the court said.

9.17   Outline of the Basic Provisions

Suppose the competitor of the local phone company was the Washington Redskins and the incumbent phone company was the Dallas Cowboys. Congress has said that the Redskins have the right to borrow Emmit Smith for any number of plays. That is called unbundling an element of the incumbent's network. And the Redskins can use the entire Cowboy team at a discount off what Jerry Jones has paid them. That is called resale. Also the Redskins can hand off the ball to Smith if their own runners aren't doing so well. That's called interconnection. If Smith helps the Redskins get a touchdown, that's called termination, for which some think the Cowboys should be paid nothing but the Skins should get the points.

(*Investor's Business Daily*, 5/20/96, A6; from Edupage)

9.18   The One Thing you Can Predict—Unpredictibility

As we work through the various components of the Act, you need to remember that this is a bird's eye view of these topics. Each one could probably be the subject of a small book in and of itself. Since regulation changes faster than technology, you need to make sure that what we tell you in this book is still valid. We know *something* has changed since we wrote this book—we just don't know what.

9.19   Watch the Sidelines—New Players May Be Coming!

As you survey the current marketplace, you might not see a niche that you can profitably fill. However, look ahead a little in the future. I anticipate that there is going to be a convergence of what we have thought of in the past as utility functions. You are already seeing many electric companies and gas companies merge. That same impulse will drive the electric companies into acquiring telephone companies. They will want to have ways to keep customers—and one way to do that is be the big one stop shop. You will be able to buy gas, electricity, telephone service, cable service, Internet access, paging, and cellular service—all from one big megacomplex.

If the electric companies get into telephone service, do you think it will be long before telephone companies decide that they can get into electricity? You will have the potential for a battle of true titans—where it takes a billion dollars just to sit down at the game. Of course, these new megaproviders will be slow and clumsy. As an entrepreneur, you should be on the lookout for ways that you can provide similar services—but faster, smarter, and better than they can.

If we do reach the day when we have huge companies providing every wire and pipe to your house, there will be a natural desire on the part of politicians to control these beasts. At that time, we will go through another round of legislative and regulatory "reform." When we do, it will be time for another book, explaining how you can find the gold in the *new* Telecommunications Act.

Please let us know if you found this book helpful—or if you found parts of it to be in error. We really do want to help you, and the best way we can do that is if you let us know what you are thinking. We are all off on a grand adventure. We will leave you with the Irish blessing,

> May the road rise to greet you. May the wind ever be at your back. May you be in Heaven an hour before the Devil knows you've died.

10
Summary and Conclusions

10.1   Becoming an "Activist" Telecommunications User

A few companies have long histories of being "activist" consumers of telecommunications services. Rather than accept a "standard" palette of service offerings, these users have exploited both the regulatory environment, as well as their economies of scale to procure telecommunications services under very favorable terms. For example, the 1984 AT&T breakup has provided the catalyst for many network advancements, including new frameworks for procuring services such as AT&T Tariff 12s, saving tens of millions of dollars over the period.

Today the Telecommunications Reform Act of 1996 is providing new opportunities to not only reduce costs but procure new services to support true 21st century business processes. Many state regulators today are also supporting innovation within the corporate community, oftentimes allowing leading edge companies to build advanced or experimental networks on a custom assembly basis. Texas Instruments, General Electric, and EDS are good examples. The network engineering organizations of other leading edge firms are exploiting this new environment, just as these companies and others exploited the old one by envisioning, designing, and developing advanced networking solutions.

USAA's first foray into this environment was successful. By exploiting changes in the regulatory environment and by judiciously utilizing outside resources with competencies in these areas, USAA Network Engineering reduced local ISDN PRI access costs in San Antonio by some $3.5 million. Due in large part to their success on this project, the department decided to look for similar success in the regions.

Why are companies like these showcased as textbook examples of innovation in telecommunications? Our rationale is simple: If *business* drives the process, the applications developed will be *usable* to business.

In order to make tomorrow's networks a reality, each "activist" user must sharpen its business, technical, and regulatory skills and become an influencer of the regulatory process itself. In this manner companies form a collaborative relationship between business, technologists, and regulators in creating new, innovative network services which equate directly to *greater growth*, *job creation*, and *competitive position*. That is good for the company, the state economy, and ultimately good for the U.S. as a whole.

I, the author, believe that a focus on job creation and growth pays large dividends to the organization in the event these networks require regulatory approval, and most of them will.

These sentiments add a "God is on our side" tenor to discussions with regulators, which is difficult to ignore. Growth, jobs, intellectual processes, growth in the state—these are all concepts that are difficult for a regulatory agency to disagree with. As long as discussions are on this high plane as opposed to the "I can save money" mode of the past, your chances for success are greatly enhanced.

If you work in a big company, the organization's size will draw considerable interest from regulators. Size alone is not all that matters these days, however. It will be necessary to combine your firm's economics of scale with a knowledge of the competitive environment to present a truly optimal solution.

So what are the opportunities available today? Recently, we have discovered a virtual plethora of opportunities, principally in the areas of unbundled network elements, dark fiber, and access arbitrage. This chapter shows how your organization should exploit these other issues after completion of your initial business analysis project.

10.2   Exploiting Telecom Reform—Defining the Project

Why isn't telecom reform "reforming" anything? Our assessment appears in this chapter, along with a strategy to push events into your company's favor in conjunction with this project. Before reading further, consider Figure 10.1. The text below will start with a discussion of the "lower hemisphere" of the pie chart, occupied by the ILECs (incumbent LECs like Bell, GTE, and others), the IXCs (interexchange carriers like AT&T), and the CLECs (competitive local exchange carriers like U.S. LEC). The lower hemisphere is characterized by the following:

10.2.1   *The Incumbent Local Exchange Carriers (ILECs)* ①[1]

The ILECs, like Bell and GTE are clearly volume businesses. Even when one considers the size of a typical business user, they are but a small part of an ILEC customer portfolio. As a result, business has not been effective at influencing pricing and services in this politically sensitive monopoly. They must accept the rates, terms, and tariffs written for a broad-based market with requirements and economies far different from theirs. The result is a nearly unanimous view by business that the ILECs are unresponsive, expensive, and inflexible. Grudgingly, USAA and companies like it have made the best they can of this less-than-optimal relationship. What choice do they have?

---

[1]In Chapter 10, all circled numbers appearing beside section headings refer to the numbered pie sections in Figure 10.1.

10.2.2   The Competitive Local Exchange Carriers (CLECs) ②

Classic CLECs are not much better because their focus is also fundamentally flawed. Rather than concentrating on innovation and new services, they are opting in as "me too" copycats of the limited palette offerings of the ILECs! Prove it to yourself. When the 10K or S1 report of just about any CLEC is reviewed, the same tired menu of services appears: local dial tone, calling cards, business phone service. Certainly the needs of "Fortune" businesses go way beyond these.

The CLECs also look lustfully at their long distance brethren in the IXC sector ③ of the chart. Our question is why? With margins on long distance often measured in the fractions of a cent, and with even the long-lumbering AT&T now lean and mean, what kinds of margins do they expect to find there? How many billboards do you see today touting four cent per minute long distance—or less?

10.2.3   The InterExchange Carriers (IXCs) ③

After nearly 15 years of cutthroat long distance competition, in the post-1984 divestiture environment the IXCs have regressed to the point where they really understand only two things well:

1. Long Distance

2. Access (only because it gets them long distance)

AT&T is still signing long-term long distance contracts, but the competition is brutal. When one receives a long distance debit card in a box of breakfast cereal, as I recently did, it is probably safe to say the market is about played out, in the traditional sense anyway. And, even though AT&T *says* they are in the local business, they really are not. To prove it, try ordering *any* local service from AT&T, but stipulate that this service will not include any potential long distance revenue, or that your company wants to pick a different long distance carrier. The result will come as no surprise.

AT&T's recent TCI acquisition may change this picture in time, but it will probably take years.

10.2.4   The Policymakers ④

The policymakers, such as the FCC and PUCs are at a loss. Imagine being in their shoes as arbitrators. The LECs want long distance access costs high. The IXCs want access costs low. The CLECs swing both ways. Everyone in the lower hemisphere of the pie is operating at such low margins, and in areas of such narrow focus. Is it any wonder why they go ballistic at any suggestion that they may lose market share to "competition?" Is it any wonder that "telecom reform" is not "reforming" anything?

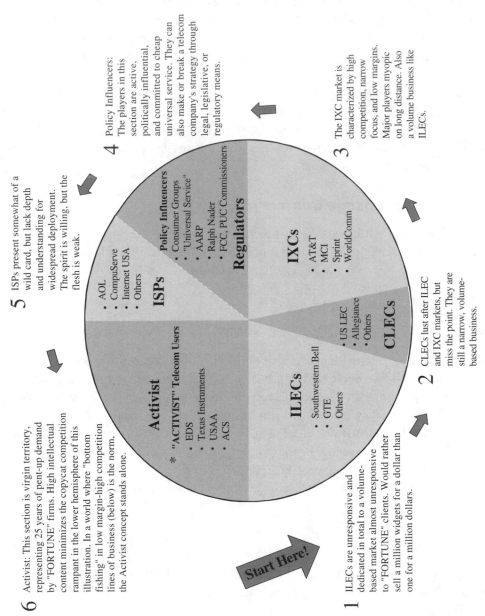

6 Activist: This section is virgin territory, representing 25 years of pent-up demand by "FORTUNE" firms. High intellectual content minimizes the copycat competition rampant in the lower hemisphere of this illustration. In a world where "bottom fishing" in low margin-high competition lines of business (below) is the norm, the Activist concept stands alone.

5 ISPs present somewhat of a wild card, but lack depth and understanding for widespread deployment. The spirit is willing, but the flesh is weak.

4 Policy Influencers: The players in this section are active, politically influential, and committed to cheap universal service. They can also make or break a telecom company's strategy through legal, legislative, or regulatory means.

3 The IXC market is characterized by high competition, narrow focus, and low margins. Major players myopic on long distance. Also a volume business like ILECs.

2 CLECs lust after ILEC and IXC markets, but miss the point. They are still a narrow, volume-based business.

1 ILECs are unresponsive and dedicated in total to a volume-based market almost unresponsive to "FORTUNE" clients. Would rather sell a million widgets for a dollar than one for a million dollars.

Start Here!

Activist

* "ACTIVIST" Telecom Users
• EDS
• Texas Instruments
• USAA
• ACS

ILECs
• Southwestern Bell
• GTE
• Others

CLECs
• US LEC
• Allegiance
• Others

IXCs
• AT&T
• MCI
• Sprint
• WorldComm

ISPs
• AOL
• CompuServe
• Internet USA
• Others

Policy Influencers
• Consumer Groups
• "Universal Service"
• AARP
• Ralph Nader
• FCC, PUC Commissioners

Regulators

* based on custom tariff filings at the Texas PUC which are part of the public record

Figure 10.1: Why "Telecom Reform" is not "reforming" anything—what can your company do?

The battleground for these three lower hemisphere occupants of the pie is the regulatory commissions. When one factors in the influence of a few very powerful consumer groups (AARP, Ralph Nader, and the like), the situation rapidly degenerates to gridlock. *Exactly the result, incidentally, that a company wishing to forestall competition desires.* This is where we have been for the last two years.

10.2.5   The Internet Service Providers (ISPs) ⑤

Why don't ISPs become CLECs? A few have. Unfortunately, ISPs do not understand telecommunications. Instead, they rely on the IXCs and LECs for solutions, which are woefully inadequate for data. The spirit is willing, but the expertise and pocketbooks will not allow true innovation anytime soon in this segment. While ISP voice offers hot prospects in the future, regulatory issues such as certification and access charges are far from decided in this area.

10.2.6   The "Activist" Fortune User ⑥

How about a different approach? Telecom reform allows "Fortune" level companies to add the intellectual component *themselves* to "wholesale parts" of telecommunications networks. These are called unbundled network elements, or UNEs. What would happen to a company that procured UNEs to build to *their* specifications and actually influenced telecom policy to its advantage? Many large firms have already by default been introduced to UNEs, and have taken an "activist" role in exploiting regulation. For example, in negotiations for "custom" access arrangements, local carriers often voluntarily "unbundle" by showing just how much the user is paying for the individual pieces, that is, fiber transport, switch ports, ancilliary features, and so on. This puts the costs in perspective and allows the buyer to negotiate based on "real" carrier costs and not just arbitrary tariff amounts. This is leading edge stuff, but very lucrative to those who pull it off.

For example, I myself, during negotiations on an ISDN PRI configuration installed for a large client in San Antonio, was successful in getting Southwestern Bell to separate "transport" from "switch port" network elements. The rationale was that my client should not pay twice (for their fiber ring, and again for PRI transport riding that ring) for transport services. That client did very well on their custom quote as a result.

There is another side to unbundling however. The Federal Telecommunications Act mandates that the local carriers unbundle network components and set the price at the LRIC (long run incremental cost) of the service—to other carriers. This is a profound change since items like dark fiber, which could possibly be the most valuable service to "Fortune" users, carry little or no LRIC cost. The catch, however, is that users would have to become carriers themselves, or dance closely with one, to avail themselves of unbundled components as specified in the Telecommunications Act. We recommend users evaluate exactly this in a future project.

A "Fortune" user in concert with a competitive local exchange carrier (CLEC) can break the regulatory gridlock in section ④ of Figure 10.1. Huge, visible networks in manufacturing, distance medicine, and finance, which create jobs, add a "God is on our side" tone to interplay with regulators unimaginable to the occupants of the lower hemisphere of the figure. Designer carriers break the "zero-sum-gain" of trading the same coconuts in sections 1, 2, and 3. They expand capacity, create jobs, and are embraced by regulators. Your company would, in this case, enjoy a state-of-the-art network, which is largely immune to "copycat" competition by competitors due to its complexity and regulatory component.

10.2.7 So, What Does This Mean?

It means everyone in the lower hemisphere of the piechart is sinking billions of dollars into capital-intensive, low-margin commodity businesses that have passed their prime. When one asks any company (CLEC, ILEC, or IXC) in the lower hemisphere what "competition" means, that company points at their target market next door—another capital-intensive, low-margin commodity business!

Virtually every local competitor competes in the lower hemisphere of the piechart. That is part of the reason telecom reform is not reforming anything. So what can your company do about it? That is reflected in the upper hemisphere. *Your organization must strive to be an "activist" telecommunications consumer and drive these processes to its own advantage.*

For the time being, your organization should consider capitalizing principally on its size and economies of scale, while making its first baby steps into the regulatory arena.

In the next step, your organization should consider an "activist" position, whereby it can manipulate telecom policy to its own advantage—opening the door to products, pricing, and management capabilities unimaginable today.

The remainder of the chapter describes both of these steps, and what we recommend your organization do in the future.

Your organization's chances for success are significantly increased when the underlying reasons for requiring a "custom deal" are communicated clearly to regulators.

10.3 Who Decides What You Pay?

Who sets the rate that your company pays for services? This is a complex question. The price you pay for services, for example, often bears no resemblance to what it costs the carrier to provide it. Consider Figure 10.2.

Some services are actually priced *below* cost. Residential dial tone is a good example. In the interest of "universal service," other servifces *subsidize* residential dial tone to keep the cost low. These services are priced above cost, and sometimes *way* above cost. The relative service contributions are shown in Figures 10.2 and 10.3. The point which should be apparent is that the services purchased by "Fortune 1000" users like Special Services

Services Priced Below Cost
Basic Telephone Service

Services Priced Above Cost
Custom Calling Features
Caller ID
Touch Tone

Services Priced *WAY* Above Cost
Private Lines
"Local" Long Distance
Digital Services

Figure 10.2: Relative service contributions.

High Contribution Services

    A. Custom Calling Features
    B. Touch Tone
    C. Special Services
    D. Digital Services
    E. "Local" Long Distance
    F. Others

Low Contribution Services

    A. Dial Tone
    B. Lifeline Services
    C. Pay Telephones

Figure 10.3: Residual ratemaking.

and Digital Services pay the highest "contribution" and carry the highest net cost. These costs are set at the state and federal level in proceedings called *Rate Cases*.

10.3.1   *Who Participates in Rate Cases?*

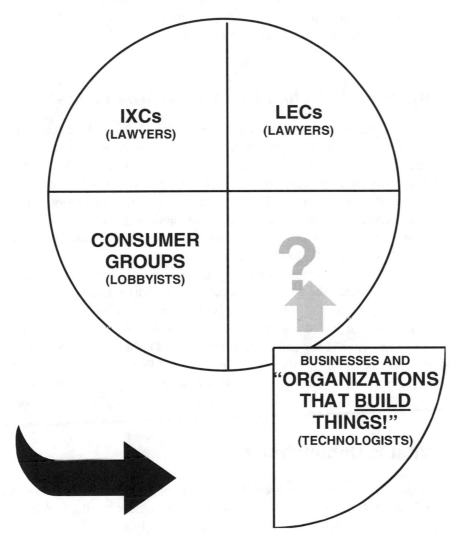

Figure 10.4: Who creates our national telecommunications policy?

Consider Figure 10.4. If you imagine the contingencies responsible for creating our national telecom policy they break into four groups, each with vastly differing goals. Starting with the lower left, and going clockwise through the figure, these include:

Contingency	Goal
1) Consumer Groups (like AARP)	Keep basic dial tone affordable
2) IXCs (like AT&T and MCI)	Keep long distance access rates *low*
3) LECs (like SBC, Bell South)	Keep long distance access rates *high*

Who do you suppose the last continent is? Why it is *you* of course—the organization which creates jobs and builds things. The only problem is, most "Fortune 1000" companies do not have direct representation in rate cases. Therefore, who do you supposed pays the highest rates?

10.3.2   What Can the Fortune User Do to Influence Rates?

You may want to consider pleading your case directly at the state public utility commission. It has been done before. Today custom pricing for long distance has become commonplace, such as AT&T's "Tariff 12." With the advent of local competition, it is becoming possible to get a "Tariff 12-like" custom tariff. Since the process is newer, it is more complicated but definitely doable. Remember, however, you will be competing for attention with the three other powerful constituencies described earlier. The public utility commission will be asking what you want to do which is worth tampering with contributions to universal service. What will your "custom tariff" do to contribute to the public wellbeing? New processes, new business solutions, and *jobs* are the best responses. A four-step process which defines business processes, defines the technology necessary to deliver them, identifies potential providers, and finally, writes the custom tariff is recommended. (See Figure 10.5.)

Also be advised that in today's environment, your firm need not pick from existing combinations of services. They can design their own. Designer Carriers™ exist which can build services to custom specifications. Dallas-based Premiere Network Services, Inc. (www.rewireit.com) is one example. A Designer Carrier™ buys *unbundled* elements and uses them as the basis for new, sophisticated services. Unbundling is explained below.

10.4   What Is Unbundling?

Imagine that the government is not breaking up the phone company. Imagine instead that they are breaking up McDonald's hamburgers. The government has decreed that McDonald's must "open" the hamburger market to potential competitors.

- McDonald's sells a Big Mac for $1.59.

- A certified competitor can purchase a Big Mac from McDonald's for $1.24—the "official" McDonald's price minus a 21.6 percent discount—and resell the finished product.

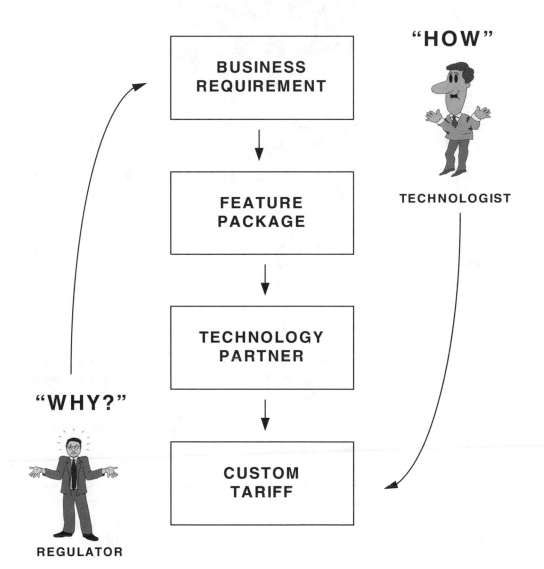

Figure 10.5: Making it work.

- The competitor can charge whatever the market bears, and is responsible for billing for the Big Mac, plus all applicable taxes.

A *facilities-based* competitor (in Texas for example) does not have to buy a Big Mac, he can buy the parts. Two all beef patties, special sauce, lettuce, cheese, and so forth. He is free to "bundle" those parts any way he wants. And here are the good parts.

- The competitor can sell the same Big Mac, by buying the individual parts *unbundled* from McDonald's (patties, lettuce, cheese, and so forth) and assembling them—only now he can build a Big Mac for 75 cents or less.

- The competitor may also find that the cheese, pickles, and condiments come as part of the price of the bun.

- Depending on the competitor's agreement, if he is a savvy negotiator, he might even get McDonald's to assemble those 75 cent Big Macs *for him*. (This is called *re-bundling* of unbundled elements.)

The following explains the dramatic advantages corporations can derive by introducing unbundled elements into their environment.

As stated previously, many firms have been successful in the past with the concept of a "custom network deal." Unlike earlier attempts at custom pricing, however, new avenues offering even more lucrative alternatives exist in the post-telecom reform environment.

Many of the unbundled components that organizations had to go through great lengths to secure in earlier projects are available now under provisions of the new federal law. Therefore, we evaluated unbundling under two scenarios, and included the results (names deleted) for your review and comparison.

- The first was under a traditional quote from the incumbent LEC (ILEC) as organizations have done in the past.

- The second was a preliminary look at how this pricing could be improved by taking advantage of specific unbundling provisions in the federal telecom law.

Both have the potential to pay back handsomely, but the new unbundling provisions could be especially appealing.

To summarize, each ILEC quote was secured with pricing broken out into the individual, "unbundled" components, just as in earlier projects (retail quote with discount). The second way we looked at unbundling was under the context of interconnection agreements available only to competitive local carriers (wholesale quote with cost-plus markup). Figure 10.6 shows (A) the OC48 tariff rate, (B) the OC48 UNE rate, and (C) the OC48 rate using "dark fiber" and a carrier interconnection agreement. Which column would *your* company like to order from?

TARIFF vs. UNE vs. DARK FIBER
SELECTED CATEGORIES OF SERVICE

	TARIFF		UNE		DARK FIBER
	monthly/installation		monthly/installation		
VG*	$110.90 /	$416.00	$63.64 /	$276.21	NONE
T1*	$593.60 /	$614.00	$256.77 /	$1,145.21	NONE
OC3*	$4,717.80 /	$821.00	$3,574.11 /	$1,301.98	$2,379.84 RETAIL***
OC12**	ICB		$16,341.78 /	$3,462.30	$9,517.35 RETAIL
OC48**	$116,000.00 /	ICB	$63,620.92 /	$3,552.36	$38,077.66 RETAIL

CONNECT POINTS: _972 (401) Northlake_ _214 (826) Taylor_ _972 (251) Irving_	* _= Point to Point (TA to NL)_ ** _= SONET Ring (TA to NL to IR)_ *** _= Includes Carrier's Margin_ _ICB = Individual Case Basis_

⇒Clearly the services of most importance such as dark fiber, carry the lowest "LRIC" cost.

⇒The question is how does your organization best exploit this opportunity?

Figure 10.6: Sample cost comparisons.

Figure 10.6 illustrates the dramatic effect of unbundling. It is part of a project we completed for another client in Dallas. It is apparent that the dynamics of unbundled "dark fiber" have changed your organization's acquisitions paradigm. Bits have gotten cheap. But first, you need to buy lots of bits.

10.5   Exploiting Today, Preparing for Tomorrow

Your organization should develop its political and regulatory experience as a "bridge" between the business community, regulators, and technologists to make the advanced technology solutions first understandable, then available. Our rationale is simple: if the *business units* drive the process, the applications developed will be *usable* to the business units. Here is a brief outline for a consultative engagement (internal or external) that will position your company to take advantage of these changes in regulation. (See Figure 10.7.)

10.5.1   Phase I—Concept Validation

In this phase you need to produce a detailed report outlining the strategy for each state in which an outlying regional office resides in order to secure optimum pricing and configurations, specifically:

- SONET metropolitan access linking AT&T, ILEC, and USAA

- SONET STS-1 handoff from AT&T and ILEC

- Accu-ring certification

- UNEs, with regard to ILEC offerings

10.5.2   Phase II—Specifications, Configurations, and Final Plans

- Final specifications by region, including final strategy, pricing, technical configuration

- Delivery of a final set of plans and specifications in a form submittable to the ILEC, utility commission, or both

- Pricing utilizing unbundling at the "retail" level—avoid paying twice for transport

10.5.3   Phase III—Regulatory Support

Obtain regulatory support, but only if required. Get a good local legal counsel in the state, but remember they are *expensive*.

10.5.4 Phase IV—Exploiting Telecommunications Act of 1996

Your company might want to consider the first three phases of this project "UNEs 101, 102, and 103," encompassing the basics of unbundling. As you can see, each one gets progressively harder. The payoff is in UNEs 104.

In the future, your company will want to use unbundled network elements in a broader sense, encompassing "800 Capture" of local 800 traffic, telecommuting, IP voice, and a host of other areas. By this time, you will be an expert in the process and will be limited only by your imagination.

"UNEs 101"

- Create specifications for required emerging technologies.
    1. SONET metropolitan access network
    2. SONET STS handoffs
    3. Accu-ring certification
- Unbundled network elements (UNEs) from retail tariffs
- Negotiate price based on TELRIC (total element long run incremental cost), taking into account the dynamics of fiber optics, and availability of UNEs.

"UNEs 102"

- Specifications, configurations, and final plans
    1. "Home" office
    2. Outlying regions

"UNEs 103"

- Regulatory support
- Custom regulatory filings (Federal and State as required)

Phase IV

- Evaluation of interconnection agreements.
- Wholesale rebundling of UNEs.
- Access arbitrage (Local 800 Capture)
- Telecommuting / Nomadic Workers / XDSL
- Nationwide SONET / ATM / IP Solutions
- Advanced Disaster Recovery Support
- Other Services

Figure 10.7: Strategy for realizing the vision.

10.6   Summary

As technologists, we are leaving no stone unturned in securing competitive quotes at the "retail" level. To best illustrate exactly where most "Fortune" users are now, consider Figure 10.8. Most firms basically have four options:

Option 1 Call, order the service, and pay the same price everyone else does. Nobody ever does this.

Option 2 Multiyear contract, with percentage off "retail" or tariffed rate. Currently used by companies but risky when price pressure is downward over the contract term.

Option 3 This is an "activist" option, but still based on a retail tariffed (and sometimes arbitrarily set) rate. This involves some kind of "partnership" and custom tariff, often negotiated in part by executives of both companies on the golf course. The key is some kind of additional pressure or influence on the service provider. This can be a Board of Director member for example.

Option 4 This option represents new ground as it utilizes a fundamentally different method of pricing. Rather than discounting off a tariff rate as in Options 1, 2, and 3, Option 4 starts with the service provider's actual wholesale cost, then works upward in a cost-plus fashion. Since components like fiber optics have extremely low LRIC costs and are the most beneficial services to large companies, you should use this methodology in future network procurements.

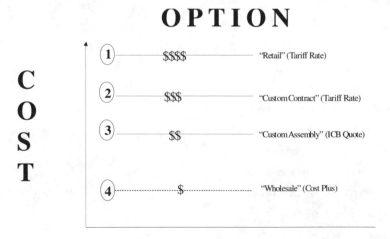

Figure 10.8: Four options.

Clearly, from the standpoint of capacity, cost, and the ability to manage capacity, the concept of buying the wholesale parts, then custom building your own "network of tomorrow" is most optimal. We hope we have piqued your interest in these new endeavors, and hope we have left you with a few tools—technical, regulatory, and managerial—you will require to bring "killer applications" together for your company, and success and notoriety to your career. Good luck in your pursuits!

A
Glossary

1010XXX Dialing 1010XXX represents the unique seven-digit access code assigned to each common carrier in the United States. Customers in Equal Access End Offices can access any carrier that purchases access into that particular LATA by dialing 1, 0, 1, 0, and the unique three-digit number assigned to the carrier selected, hence the term 1010XXX.

Access Tandem The primary serving office for a given LATA through which much of the common carrier access is derived. Some types of access links like Feature Group "B" are provided almost exclusively from the Access Tandem. Others such as Feature Group "D" may be derived from the Access Tandem or an Equal Access End Office.

"Backhoe Fade" Slang term common among telecommunications professionals, meaning "cable cut."

CLEC Competitive Local Exchange Carrier.

Cell Site A cellular radio transmitter that provides mobile telephone service to a particular area of town. Mobile telephone users have their calls switched from cell to cell as they drive from one part of town to another, otherwise known as a "hand off" between cell sites.

Competitive Access Provider (CAP) Any non-Bell network within a metropolitan area which has the purpose of providing local exchange services similar to those offered by the Bell Operating Companies. Sometimes called "alternate access" carriers.

DCS Digital Crossconnect System. Commonly used in central offices for cross connecting circuits. DACS units can connect T1s, individual circuits, or elaborate combinations of the two.

Demark Common telephone company term for the telephone company's network interface installed at the customer location. Telephone company responsibility starts and stops at this location, with inside wiring now the responsibility of the customer and not the telephone company.

Digital Hub Designated central office which provides maintenance, circuit timing, and other advanced functions for digital data circuits. While more offices now provide digital services, in many cases "premium" digital service routes through this specialized office.

DISA Direct Inward System Access. Allows the user the ability to dial into a PBX, receive a dial tone, enter an access code, and complete calls on the company network. These systems are convenient but subject to high levels of fraud and abuse by unauthorized users.

Divestiture The 1984 breakup of AT&T into AT&T Communications, AT&T Bell Labs, and AT&T Technologies which also created the seven regional Bell Telephone Companies such as NYNEX and Bell South.

DS-1 Another term for T1, or T-Carrier. A 1.544 Mbps circuit.

DS-3 Another term for a T3 carrier. A 45 Mbps circuit.

End Office The type of central office that provides local exchange services directly to end users, as opposed to other central offices. Also called the Local Serving Office.

Fiber Optic Loops Term denotes redundant paths built into fiber optic cable routes to automatically or manually divert traffic in the event of failure of the primary path.

HALON Fire Systems Commonly used in computer rooms, HALON is an odorless gas which is discharged quickly into a room to extinguish a fire by cutting off oxygen, while still leaving enough to allow for respiration of the room's occupants. It is generally considered harmless to humans, although it is expensive and sometimes subject to accidental discharge.

Hitless Switching Refers to a technology applied to many communications mediums (microwave, fiber, and so forth) to provide a very fast "switch" to a backup facility in the event of failure on a primary path. This happens quickly enough to assure that the facility does not drop calls already in progress, or disrupt data, hence the term "hitless."

IntraLATA Refers to circuits that originate and terminate in the Local Access Transport Area, which is serviced by the local (Bell or GTE) operating company.

InterLATA Refers to circuits that originate in one Local Access Transport Area, but terminate in a different LATA. These circuits are carried by AT&T or another long distance company.

ISDN Integrated Services Digital Network. An advanced digital service offering designed as an industry standard to provide much greater capacity and more advanced features than standard telephone service.

Local Access Cable The cable facility from the Local Serving Office to the end user. The last link of a circuit to the user.

Local Area Network (LAN). Characterized as a network serving a bounded area, normally within a building or at most a campus environment, operating using various platforms. The most popular include Ethernet, Token Ring, Arcnet, and so forth.

Local Serving Office The serving office from which the user draws regular dial telephone services. It is also the last central office utilized before a circuit is connected to an end user.

Loop The copper facility connecting a local serving office with the subscriber. For example, "copper loop."

M24/M44 Multiplexer Term that denotes a central office D4-compatible T1 multiplexer which conforms to AT&T's M24 or M44 standard. M24 multiplexers combine 24–64 Kbps channels into a single T1. M44 multiplexers use ADPCM multiplexing to derive 44–32 Kbps channels from a T1 circuit.

NXX Codes An NXX code is the first three digits of a telephone number which identify the central office to which the number is assigned. By using the area code and NXX code (such as 617-223-XXXX), central offices anywhere in North America can be readily identified.

OCC Term common in the predivestiture era and still used today. It stands for "Other Common Carrier" (other than AT&T). The proper term is IXC, Interexchange Carrier.

POP Point of Presence. A POP is a meeting place where local exchange facilities connect to long distance companies. Hence the terms, AT&T POP, MCI POP, and so on.

Private Line Circuit Also called a "special" or "custom" circuit. The circuit can be for voice or data. Private line circuits provide any service which is anything other than normal dial tone service.

Right of Way The public utility easement through which public services such as water, sewer, telephone, cable television, electricity, and other services are routed.

Robust Network A term finding increasing use, particularly among Bell operating companies and AT&T. It describes a network with fault-tolerant systems designed to be resistant to accidental service disruptions.

RS-232 A 25-pin computer to modem connector.

Special Facility Construction Broad term given to any type of unique, custom services constructed by the Bell operating company specifically for a large customer.

T1 A 1.544 Mbps circuit that provides 24 64-Kbps channels.

T3 A circuit which carries 28 T1s (672 channels) and operates at the speed of 44.736 Mbps. T3s were originally used solely for voice, linking AT&T "Class 1" central offices in the 1970s. They are used for a variety of other purposes today including data and video.

T-Carrier Term commonly used within the Bell operating companies for T1 circuits installed between Bell serving offices.

Transponder A transponder is a single "channel" on a satellite. There are 24 transponders on each satellite.

Wide Area Network Generally accepted as the "telco" portion of the network, encompassing local and long distance telephone companies, and characterized by the fact that it covers an area of many miles, as opposed to Local Area Network (LAN), Campus Area Network (CAN), Metropolitan Area Network (MAN), and other such terms.

B
Common Telecommunications Acronyms

Acronym	Description
A/D	Analog to Digital conversion
A/T	Access/Tandem
AAB	Automatic Answer Back
AABS	Automated Alternate Billing System (system through which O+ calls are billed)
AAP	Alternate Access Provider
ABS	Alternate Billing Service
ACH	Automated Clearinghouse
ACN	Access Customer Name
ACN	Advanced Communication Network
ACNA	Access Carrier Name Abbreviation (Bellcore assigned)
ACR	Anonymous Call Rejection
ACRJ	Automatic Caller Rejection
ACTEL	Access Telephone Number
ACTL	Access Customer Terminal Location
ADL	Additional Directory Listing
ADN	Advanced Digital Network
ADVC	Advanced/Payment/Deposit Verification
AECN	Alternate Exchange Carrier Number (SWBT term synonymous to an OCN)
AF	Alternate Final (overflow trunk in E9-1-1 Service)
AFR	Automatic Flexible Routing
AIN	Advanced Intelligence Network
AIN-SCP	Advanced Intelligence Network—Service Control Point
AIOD/IOD	Automatically Identified Outward Dialing/Identified Outward Dialing
AIS	Automatic Intercept Service
ALI	Automatic Location Identification System (E911 service)

ALIS	Automatic Location Identification System
ALT	Automatic Line Testing
AMI	Alternate Marked Version
AML	Additional Main Listing
ANI	Automatic Number Identification
ANSI	American National Standards Institute
AP	Access Provider
APOT	Additional Point Of Termination
ARS	Automatic Route Selection
ARSB	Automated Repair Service Bureau
ASCII	American Standard Code for Information Exchange
ASR	Access Service Request
ASYNC	Asynchronous
AT	Access Tandem
ATC	Access Tandem Code
B8ZS	Bipolar 8 with Zero Substitution
BAUD	A unit of signaling speed (one baud corresponds to a rate of one signal element per second)
BBS	Bulletin Board System
BDT	Billing Data Tape
BER	Bit Error Rate
BES	Basic Exchange Services
BLV	Busy Line Verification Code (trunk groups)
BLV/BLI	Busy Line Verify/Busy Line Interrupt
BLVI	Busy Line Verification & Intercept
BN	Billing Number
BNA	Bill Name & Address
BNS	Billed Number Screening
BOC	Bell Operating Company
BOCP	Billing Optional Calling Plans or Plus
BOCP	Business Optional Calling Plan
BRI	Basic Rate Interface
Bellcore	Bell Communications Research
CABS	Carrier Access Billing System
CADS	Computer Aided Dispatch System (E9-1-1)
CAPS	Competitive Access Providers
CCCC	Customer Controlled Cross Connect
CCECC	Co-Carrier Equipment Cross Connect (also called "cage to cage")

CCL	Carrier Common Line
CCLC	Carrier Common Line Charge
CCN	Certificate of Convenience & Necessity
CCON	Customer Contact
CCRS	Centrex Customer Rearrangement System
CCS	Common Channel Signaling
CCS7	Common Channel Signaling System 7
CCSA	Common Channel Signaling Access Capability (provides SS7 signaling for FG-D)
CCW	Cancel Call Waiting
CENTREX	Central Office Exchange Service
CFB	Call Forwarding Busy
CFBL	Call Forwarding Busy Line
CFW	Call Forwarding
CIC	Carrier Identification Code
CIR	Circular Hunting
CKID	Circuit Identification
CKL	Circuit Location
CKLT	Circuit Location (Central Office interface point bridging location)
CKLT	Circuit Location Telephone Company Wire Center
CKTID	Circuit Identification
CLASS	Customer Local Area Switching Services
CLEC	Competitive Local Exchange Carrier
CLID	Calling Line ID Display
CLLI	Common Language Location Identification
CLSV	Class of Service
CMRS	Commercial Mobile Radio Service
CMS	Centrex Management Service
CNAM	Caller ID Name
CNB	Call Number Blocking
CND	Calling Number Delivery
CNDB	Calling Number Delivery Blocking
CNDS,SWBIC	Calling Name Delivery Service
CNI	Common Network Interface
CNNB	Calling Number Blocking
CNND	Calling Number Delivery
CNR	Customer Not Ready

CO	Central Office
COC	Central Office Code
COPT	Customer Owned Pay Telephone (now called PSP [Payphone Service Provider])
COPW	Customer Owned Premises Wire
CPAC	Circuit Provisioning Assignment Center
CPAT	Competitive Provider Account Team
CPCN	Certificate of Public Convenience and Necessity
CPE	Customer Premises Equipment
CPG	Call Pickup Group
CPIW	Customer Provided Inside Wire
CPM	Cost Per Minute
CPN	Calling Party Number
CPNI	Customer Proprietary Network Information
CPP/PPP	Calling Party Pays/Paging Party Pays (Wireless term)
CPU	Call Pick Up
CR	Call Return
CRDD	Customer Requested Due Date
CRIS	Customer Record Information System
CRS	Cell Relay Service
CS	Call Screen
CS	Class of Service
CSN	Call Screening
CSN	Customer Service Node
CSR	Customer Service Record
CSU/DSU	Channel Service Units/Data Service Units
CT	Call Trace
CTC	Call Trace Center
CWG	Call Waiting
CWI	Call Waiting Indicator
CWL	Call Waiting Lamp
CXR	Call Transfer
DA	Directory Assistance
DAC	Directory Assistance Charge
DACC	Directory Assistance Call Completion
DACS	Digital Access Connect System
DCF	Delay Call Forwarding
DCS	Digital Cross Connect

DDD	Direct Distance Dialing
DDI	Direct Digital Interface service (Centrex)
DDN	Dialable Directory Number
DDOV	Digital Data Over Voice Service
DDS	Digital Data Service
DID	Direct Inward Dial
DLC	Digital Loop Carrier
DLR	Design Layout Reports
DNIS	Dialed Number Identification Service
DNR	Do Not Refer
DNRI	Directory Number Route Index
DOD	Direct Outward Dial
DS-0	Digital Signal Level 0
DS-1	Digital Signal Level 1
DSU/CSU	Data Service Unit/Channel Service Unit (Customer interface for DDS channel bank)
DSX	Digital Cross Connect
E9-1-1	Emergency 911 database
EA	Equal Access
EAEO	Equal Access End Office
EAS	Extended Area Service
EO	End Office/End Office switch
ETN	Electronic Tandem Network
ETR	Estimated Time of Restoral
ETTR	Estimated Time To Repair
EUCL	End User Common Line Charge
FCC	Federal Communications Commission
FG	Feature Group
FG-A	Feature Group A - Line side connection from P*B
FG-B	Feature Group B - Trunk side connection from a P*B
FG-D	Feature Group D - Switched access product
FR	Flat Rate
FSO	Foreign Service Office
FTA	Federal Telecommunication Act
FTP	File Transfer Protocol
GS	Ground Start
HDSL	High bit-rate Digital Subscriber Line
HiCAP	High Capacity Service

ILEC	Incumbent Local Exchange Carrier
IP	Internet Protocol
ISDN	Integrated Services Digital Network ISDN Service Center:
Kbps	Kilobits per second
kHz	Kilohertz
KTS	Key Telephone System
LIDB	Line Information Database
LOA	Letter of Agency
LOA	Letter of Authorization (from end user)
LPIC	Local Primary IntraLATA Carrier
LSO	Local Serving Office
LSR	Local Service Request
Mbps	Megabits per second
MTS	Message Telephone Service
MTSO	Mobile Telephone Switching Office
NT1	Network Termination Device (ISDN)
PIC	Predesignated Interexchange Carrier
PIC	Presubscription Indicator
PIC	Primary Interexchange Carrier
POI	Point Of Interconnection
POT	Point Of Termination
POTS	Plain Old Telephone Service
PSAP	Public Safety Answering Point
PSN	Public Switched Network
PSTN	Public Switched Telephone Network
PUC	Public Utilities Commission
PURA	Texas Public Utility Regulatory Act of 1995
PVC	Permanent Virtual Circuit (PL)
RBOC	Regional Bell Operating Company
RFP	Request For Proposal
SMDR	Station Message Detail Recording
SONET	Synchronous Optical Network
SPOC	Single Point Of Contact
T-1	(DS1 Loops) Provides a 4-wire facility capable of transporting 1.544 Mbps DS1 digital service between two 4-wire interfaces
TDM	Time Division Multiplexing (HI-CAPS)
TN	Telephone Number
VF	Voice Frequency

VTAM	Virtual Telecommunications Access Method
WATS	Wide Area Telecommunications Service
x-CONN	Cross Connection
ZBTSI	Zero Bit Time Switch Interface

Biography

Leo A. Wrobel has more than two decades of experience in emerging network technology, disaster recovery planning, and technical training. Wrobel, a nationally renowned technical futurist, has been President and CEO of the Dallas-based telecommunications firm, Premiere Network Services Inc., since 1986.

An active author and lecturer, Wrobel has published over nine books and dozens of trade articles on a variety of technical subjects including such noteworthy works as the following:

- *Managing Emerging Technologies for Strategic Advantage* (Computer Economics)

- *Writing Disaster Recovery Plans for Telecom & LANs* (Artech House)

- *Disaster Recovery Planning for Telecommunications* (Artech House)

- *Implementing a Successful Telecommunications Disaster Recovery Plan* (Computer Economics)

- *Business Resumption Planning* (Computer Economics)

- *Understanding Emerging Network Services, Pricing and Regulation* (Artech House)

Wrobel, a knowledgeable and effective communicator known for his entertaining presentation style, lectures and teaches on a wide variety of technical topics. Demand for his speaking talent has taken him throughout the United States and overseas to Israel, South America, and other locations. Previously, a three-term City Councilman and elected Mayor, Wrobel is a skilled and powerful communicator known for his ability to build consensus.

Wrobel's experience includes assignments at AT&T, a major mortgage banking company, and a host of other firms engaged in banking, brokerage, heavy manufacturing, telecommunications services, and government. His achievements have included the design and regulatory approval of a LATA-wide OC12/ATM network for a $10 billion manufacturing giant, the first of its kind anywhere.

Wrobel holds degrees in Telecommunications Systems Technology, Electronic Systems Technology, and Business and Public Policy.

Questions can be directed to Leo Wrobel at his company's Web site, www.rewireit.com, or by calling 1-888-REWIREIT.

IEEE
COMPUTER
SOCIETY

Press Activities Board

IEEE Computer Society Publications

The world-renowned IEEE Computer Society publishes, promotes, and distributes a wide variety of authoritative computer science and engineering texts. These books are available from most retail outlets. Visit the Online Catalog, *http://computer.org*, for a list of products.

IEEE Computer Society Proceedings

The IEEE Computer Society also produces and actively promotes the proceedings of more than 141 acclaimed international conferences each year in multimedia formats that include hard and softcover books, CD-ROMs, videos, and on-line publications.

For information on the IEEE Computer Society proceedings, send e-mail to cs.books@computer.org or write to Proceedings, IEEE Computer Society, P.O. Box 3014, 10662 Los Vaqueros Circle, Los Alamitos, CA 90720-1314. Telephone +1 714-821-8380. FAX +1 714-761-1784.

Additional information regarding the Computer Society, conferences and proceedings, CD-ROMs, videos, and books can also be accessed from our web site at *http://computer.org/cspress*